Communications in Computer and Information Science   1573

More information about this series at https://link.springer.com/bookseries/7899

Evgeny Burnaev · Dmitry I. Ignatov ·
Sergei Ivanov et al. (Eds.)

Recent Trends in Analysis of Images, Social Networks and Texts

10th International Conference, AIST 2021
Tbilisi, Georgia, December 16–18, 2021
Revised Selected Papers

Springer

For the full list of editors *see next page*

ISSN 1865-0929 ISSN 1865-0937 (electronic)
Communications in Computer and Information Science
ISBN 978-3-031-15167-5 ISBN 978-3-031-15168-2 (eBook)
https://doi.org/10.1007/978-3-031-15168-2

This Springer imprint is published by the registered company Springer Nature Switzerland AG
The registered company address is: Gewerbestrasse 11, 6330 Cham, Switzerland

Editors

Evgeny Burnaev (iD)
Skolkovo Institute of Science and Technology
Moscow, Russia

Sergei Ivanov
Skolkovo Institute of Science and Technology
Moscow, Russia

Olessia Koltsova (iD)
National Research University Higher School
of Economics
St. Petersburg, Russia

Sergei O. Kuznetsov (iD)
National Research University Higher School
of Economics
Moscow, Russia

Amedeo Napoli (iD)
LORIA
Vandœuvre lès Nancy, France

Panos M. Pardalos (iD)
Industrial and Systems Engineering
University of Florida
Gainesville, USA

Andrey V. Savchenko (iD)
National Research University Higher School
of Economics
Nizhny Novgorod, Russia

Elena Tutubalina (iD)
Kazan Federal University
Kazan, Russia

Dmitry I. Ignatov (iD)
National Research University Higher School
of Economics
Moscow, Russia

Michael Khachay (iD)
Krasovskii Institute of Mathematics
and Mechanics of Russian Academy
of Sciences
Yekaterinburg, Russia

Andrei Kutuzov (iD)
University of Oslo
Oslo, Norway

Natalia Loukachevitch (iD)
Research Computing Center
Lomonosov Moscow State University
Moscow, Russia

Alexander Panchenko (iD)
Skolkovo Institute of Science and Technology
Moscow, Russia

Jari Saramäki (iD)
Aalto University
Espoo, Finland

Evgenii Tsymbalov (iD)
Yandex LLC
Moscow, Russia

Preface

This volume contains the refereed proceedings of the 10th International Conference on Analysis of Images, Social Networks, and Texts (AIST 2021)[1]. The previous conferences (during 2012–2020) attracted a significant number of data scientists, students, researchers, academics, and engineers working on interdisciplinary data analysis of images, texts, and social networks. The broad scope of AIST makes it an event where researchers from different domains, such as image and text processing, exploiting various data analysis techniques, can meet and exchange ideas. As the test of time has shown, this leads to the cross-fertilisation of ideas between researchers relying on modern data analysis machinery.

Therefore, AIST 2021 brought together all kinds of applications of data mining and machine learning techniques. The conference allowed specialists from different fields to meet each other, present their work, and discuss both theoretical and practical aspects of their data analysis problems. Another important aim of the conference was to stimulate scientists and people from industry to benefit from knowledge exchange and identify possible grounds for fruitful collaboration.

The conference was held during December 16–18, 2021. The conference was organized in a hybrid mode by Ivane Javakhishvili Tbilisi State University, Georgia (offline on campus)[2] and the Skolkovo Institute of Science and Technology, Russia (online)[3], due to the COVID-19 pandemic constraints.

This year, the key topics of AIST were grouped into five tracks:

1. Data Analysis and Machine Learning chaired by Sergei O. Kuznetsov (HSE University, Russia), Amedeo Napoli (Loria, France), and Evgenii Tsymbalov (Yandex, Russia)
2. Natural Language Processing chaired by Natalia Loukachevitch (Lomonosov Moscow State University, Russia), Andrey Kutuzov (University of Oslo, Norway), and Elena Tutubalina (Kazan Federal University and HSE University, Russia)
3. Social Network Analysis chaired by Sergei Ivanov (Huawei), Olessia Koltsova (HSE University, Russia), and Jari Saramäki (Aalto University, Finland)
4. Computer Vision chaired by Evgeny Burnaev (Skolkovo Institute of Science and Technology, Russia) and Andrey V. Savchenko (HSE University, Russia)
5. Theoretical Machine Learning and Optimization chaired by Panos M. Pardalos (University of Florida, USA) and Michael Khachay (IMM UB RAS and Ural Federal University, Russia)

The Program Committee and the reviewers of the conference included 128 well-known experts in data mining and machine learning, natural language processing, image

---

[1] https://aistconf.org.

[2] https://www.tsu.ge/en.

[3] https://www.skoltech.ru/en.

processing, social network analysis, and related areas from leading institutions of many countries including Australia, Austria, the Czech Republic, France, Germany, Greece, India, Iran, Ireland, Italy, Japan, Lithuania, Norway, Qatar, Romania, Russia, Slovenia, Spain, Taiwan, Ukraine, the UK, and the USA. This year, we received 118 submissions, mostly from Russia but also from Algeria, Brazil, Finland, Germany, India, Norway, Pakistan, Serbia, Spain, Ukraine, the UK, and the USA.

Out of the 118 submissions, 26 were desk rejected. For the remaining 92 papers, only 25 were accepted into the main volume published in Springer's Lecture Notes in Computer Science (LNCS) series. In order to encourage young practitioners and researchers, we have included 17 papers in this companion volume in the Communications in Computer and Information Science (CCIS) series. Thus, the acceptance rate of the CCIS volume is 25%. Each submission was reviewed by at least three reviewers, experts in their fields, in order to supply detailed and helpful comments.

The conference featured several invited talks and tutorials dedicated to current trends and challenges in the respective areas.

The invited talks from academia were on natural language processing and related problems:

- Magda Tsintsadze, Manana Khachidze, and Maia Archuadze (Tbilisi State University): "On Georgian Text Processing Toolkit Development"
- Jeremy Barnes (University of the Basque Country): "Is it time to move beyond sentence classification?"
- Irina Nikishina (Skolkovo Institute of Science and Technology): "Taxonomy Enrichment with Text and Graph Vector Representations"
- Zulfat Miftahudinov (Kazan Federal University and Insilico Medicine): "Drug and Disease Interpretation Learning with Biomedical Entity Representation Transformer"

The invited industry speakers gave the following talks:

- Iosif Itkin (Exactpro): "Data intensive software testing"
- Aleksandr Semenov (Sber.Games): "Data science in GameDev"
- Alexey Drutsa (Yandex): "Toloka: professional hands-on tools accelerating the data-centric AI"

This year the program also included two round tables. The first one was devoted to the largest AIST track: "NLP on AIST" chaired by Andrey Kutuzov. The second one facilitated discussion on social aspects of data science practices: "Ethical issues and social challenges in data science" chaired by Rostislav Yavorskiy.

We would like to thank the authors for submitting their papers and the members of the Program Committee for their efforts in providing exhaustive reviews.

We would also like to express our special gratitude to all the invited speakers and industry representatives. We deeply thank all the partners and sponsors, especially, the hosting organization: Ivane Javakhishvili Tbilisi State University in Georgia. The local organizing team also included representatives of Exactpro Systems: Iosif Itkin, Natia Sirbiladze, and Janna Zabolotnaya. In addition, we thank the co-organizers from the Russian side: Alexander Panchenko and Irina Nikishina of Skolkovo Institute of Science and Technology and the colleagues from various divisions of HSE University. Our special thanks go to Springer for their help, starting from the first conference call to the final version of the proceedings. Last but not least, we are grateful to the volunteers, whose endless energy saved us at the most critical stages of the conference preparation.

Here, we would like to mention that the Russian word "aist" is more than just a simple abbreviation as (in Cyrillic) it means "stork". Since it is a wonderful free bird, a symbol of happiness and peace, this stork gave us the inspiration to organize the AIST conference series. So we believe that this conference will still likewise bring inspiration to data scientists around the world!

December 2021

Evgeny Burnaev
Dmitry I. Ignatov
Sergei Ivanov
Michael Khachay
Olessia Koltsova
Andrey Kutuzov
Sergei O. Kuznetsov
Natalia Loukachevitch
Amedeo Napoli
Panos M. Pardalos
Jari Saramäki
Andrey Savchenko
Evgenii Tsymbalov
Elena Tutubalina

Organization

The conference was organized by a joint team from the Ivane Javakhishvili Tbilisi State University and Exactpro Systems in Georgia, the Skolkovo Institute of Science and Technology (Skoltech), various divisions of the National Research University Higher School of Economics (HSE University), and the Krasovskii Institute of Mathematics and Mechanics of the Russian Academy of Sciences (IMM UB RAS).

Program Committee Chairs

Evgeny Burnaev	Skolkovo Institute of Science and Technology, Russia
Sergei Ivanov	.
Michael Khachay	IMM UB RAS and Ural Federal University, Russia
Olessia Koltsova	HSE University, St. Petersburg, Russia
Andrey Kutuzov	University of Oslo, Norway
Sergei Kuznetsov	HSE University, Moscow, Russia
Natalia Loukachevitch	Lomonosov Moscow State University, Russia
Amedeo Napoli	Loria, CNRS, University of Lorraine, and Inria, France
Panos Pardalos	University of Florida, USA
Jari Saramäki	Aalto University, Finland
Andrey Savchenko	HSE University, Nizhny Novgorod, Russia
Elena Tutubalina	HSE University, Moscow, and Kazan Federal University, Russia
Evgenii Tsymbalov	Yandex, Russia

Proceedings Chair

Dmitry I. Ignatov	HSE University, Moscow, Russia

Steering Committee

Dmitry I. Ignatov	HSE University, Moscow, Russia
Michael Khachay	IMM UB RAS and Ural Federal University, Russia
Alexander Panchenko	Skolkovo Institute of Science and Technology, Russia
Andrey Savchenko	HSE University, Nizhny Novgorod, Russia
Rostislav Yavorskiy	Tomsk Polytechnic University, Russia

Program Committee

Anton Alekseev	St. Petersburg Department of V.A.Steklov Institute of Mathematics, RAS, Russia
Ilseyar Alimova	Kazan Federal University, Russia
Vladimir Arlazarov	Smart Engines Service LLC and Federal Research Center "Computer Science and Control" of RAS, Russia
Aleksey Artamonov	Yandex, Russia
Ekaterina Artemova	HSE University, Russia
Aleksandr Babii	HSE University, Russia
Yulia Badryzlova	HSE University, Russia
Jaume Baixeries	Universitat Politècnica de Catalunya, Spain
Amir Bakarov	HSE University, Russia
Artem Baklanov	International Institute for Applied Systems Analysis, Austria
Nikita Basov	St. Petersburg State University, Russia
Vladimir Batagelj	University of Ljubljana, Slovenia
Tatiana Batura	Ershov Institute of Informatics Systems, SB RAS, and Novosibirsk State University, Russia
Malay Bhattacharyya	Indian Statistical Institute, Kolkata, India
Mikhail Bogatyrev	Tula State University, Russia
Elena Bolshakova	Moscow State Lomonosov University, Russia
Evgeny Burnaev	Skolkovo Institute of Science and Technology, Russia
Aleksey Buzmakov	HSE University, Russia
Mikhail Chernoskutov	IMM UB RAS and Ural Federal University, Russia
Alexey Chernyavskiy	Philips Innovation Labs, Russia
Daniil Chernyshev	Lomonosov Moscow State University, Russia
Vera Davydova	Sber AI, Russia
Boris Dobrov	Lomonosov Moscow State University, Russia
Ivan Drokin	Botkin.ai, Russia
Anton Eremeev	Omsk Branch of Sobolev Institute of Mathematics, SB RAS, Russia
Elena Ericheva	Botkin.ai, Russia
Anna Ermolayeva	Peoples' Friendship University of Russia, Russia
Adil Erzin	Sobolev Institute of Mathematics, SB RAS, Russia
Alena Fenogenova	SberDevices, Russia
Elena Filatova	City University of New York, USA
Yuriy Gapanyuk	Bauman Moscow State Technical University, Russia

Petr Gladilin	ITMO University, Russia
Maksim Glazkov	Neuro.net, USA
Anna Glazkova	University of Tyumen, Russia
Taisia Glushkova	Instituto de Telecomunicações, Portugal
Natalia Grabar	Université de Lille, France
Dmitry Granovsky	Yandex, Russia
Dmitry Ignatov	HSE University, Russia
Dmitry Ilvovsky	HSE University, Russia
Max Ionov	Goethe University Frankfurt, Germany, and Moscow State University, Russia
Sergei Ivanov	Skolkovo Institute of Science and Technology, Russia
Vladimir Ivanov	Innopolis University, Russia
Ilia Karpov	HSE University, Russia
Yury Kashnitsky	Elsevier, The Netherlands
Alexander Kazakov	Matrosov Institute for System Dynamics and Control Theory, SB RAS, Russia
Michael Khachay	IMM UB RAS and Ural Federal University, Russia
Javad Khodadoust	Payame Noor University, Iran
Denis Kirjanov	HSE University, Russia
Yury Kochetov	Sobolev Institute of Mathematics, SB RAS, Russia
Sergei Koltcov	HSE University, Russia
Olessia Koltsova	HSE University, Russia
Jan Konecny	Palacký University Olomouc, Czech Republic
Anton Konushin	Samsung and HSE University, Russia
Andrey Kopylov	Tula State University, Russia
Evgeny Kotelnikov	Vyatka State University, Russia
Ekaterina Krekhovets	HSE University, Russia
Angelina Kudriavtseva	Lomonosov Moscow State University, Russia
Sofya Kulikova	HSE University, Russia
Maria Kunilovskaya	University of Wolverhampton, UK
Anvar Kurmukov	Kharkevich Institute for Information Transmission Problems of RAS, Russia
Andrey Kutuzov	University of Oslo, Norway
Elizaveta Kuzmenko	University of Trento, Italy
Andrey Kuznetsov	Samara National Research University, Russia
Sergei Kuznetsov	HSE University, Russia
Dmitri Kvasov	University of Calabria, Italy
Florence Le Ber	Université de Strasbourg, France
Anna Lempert	Matrosov Institute for System Dynamics and Control Theory, SB RAS, Russia

Vlad Shakhuro	Lomonosov Moscow State University, Russia
Tatiana Shavrina	HSE University, Russia
Andrey Shcherbakov	University of Melbourne, Australia
Henry Soldano	Laboratoire d'Informatique de Paris Nord, France
Alexey Sorokin	Moscow State University, Russia
Andrey Sozykin	IMM UB RAS, Russia
Dmitry Stepanov	Program System Institute of RAS, Russia
Vadim Strijov	Moscow Institute of Physics and Technology, Russia
Tatiana Tchemisova	University of Aveiro, Portugal
Irina Temnikova	Qatar Computing Research Institute, Qatar
Mikhail Tikhomirov	Moscow State University, Russia
Martin Trnecka	Palacký University Olomouc, Czech Republic
Yuliya Trofimova	HSE University, Russia
Christos Tryfonopoulos	University of Peloponnese, Greece
Evgenii Tsymbalov	Yandex, Russia
Elena Tutubalina	HSE University, Russia
Valery Volokha	ITMO University, Russia
Konstantin Vorontsov	FORECSYS and Moscow Institute of Physics and Technology, Russia
Ekaterina Vylomova	The University of Melbourne, Australia
Dmitry Yashunin	HARMAN International, USA
Alexey Zaytsev	Skolkovo Institute of Science and Technology, Russia
Nikolai Zolotykh	University of Nizhni Novgorod, Russia

Organizing Committee

Dmitry Ignatov	HSE University, Moscow, Russia
Alexander Panchenko	Skolkovo Institute of Science and Technology, Russia
Irina Nikishina	Skolkovo Institute of Science and Technology, Russia
Iosif Itkin	Exactpro Systems, Georgia
Natia Sirbiladze	Exactpro Systems, Georgia
Janna Zabolotnaya	Exactpro Systems, Georgia
Alexandra Dukhaniana	Skolkovo Institute of Science and Technology, Russia

Organizing Institutions

Ivane Javakhishvili Tbilisi State University, Georgia
Skolkovo Institute of Science and Technology, Russia

Krasovskii Institute of Mathematics and Mechanics, Ural Branch of the Russian
    Academy of Sciences, Russia
School of Data Analysis and Artificial Intelligence, HSE University, Moscow, Russia
Laboratory of Algorithms and Technologies for Networks Analysis, HSE University,
    Nizhny Novgorod, Russia
International Laboratory for Applied Network Research, HSE University, Moscow,
    Russia
Laboratory for Models and Methods of Computational Pragmatics, HSE University,
    Moscow, Russia
Laboratory for Social and Cognitive Informatics, HSE University, St. Petersburg,
    Russia
Research Group "Machine Learning on Graphs", HSE University, Moscow, Russia

Additional Reviewers

Deligiannis, Kimon
Dokuka, Sofia
Fedyanin, Kirill
Florinsky, Mikhail
Koloveas, Paris
Neznakhina, Ekaterina
Ogorodnikov, Yuri
Pugachev, Alexander
Vytovtov, Petr

Sponsoring Institutions

Boeing
Exactpro Systems, Georgia
HSE University, Russia
Ivane Javakhishvili Tbilisi State University, Georgia
Skolkovo Institute of Science and Technology, Russia

Contents

Data Analysis and Machine Learning

Social Network Analysis

Theoretical Machine Learning and Optimization

Natural Language Processing

Call Larisa Ivanovna: Code-Switching Fools Multilingual NLU Models

Alexey Birshert[✉] and Ekaterina Artemova

National Research University Higher School of Economics, Moscow, Russia
{abirshert,elartemova}@hse.ru
https://cs.hse.ru/en/ai/computational-pragmatics/

Abstract. Practical needs of developing task-oriented dialogue assistants require the ability to understand many languages. Novel benchmarks for multilingual natural language understanding (NLU) include monolingual sentences in several languages, annotated with intents and slots. In such setup models for cross-lingual transfer show remarkable performance in joint intent recognition and slot filling. However, existing benchmarks lack of code-switched utterances, which are difficult to gather and label due to complexity in the grammatical structure. The evaluation of NLU models seems biased and limited, since code-switching is being left out of scope.

Our work adopts recognized methods to generate plausible and naturally-sounding code-switched utterances and uses them to create a synthetic code-switched test set. Based on experiments, we report that the state-of-the-art NLU models are unable to handle code-switching. At worst, the performance, evaluated by semantic accuracy, drops as low as 15% from 80% across languages. Further we show, that pre-training on synthetic code-mixed data helps to maintain performance on the proposed test set at a comparable level with monolingual data. Finally, we analyze different language pairs and show that the closer the languages are, the better the NLU model handles their alternation. This is in line with the common understanding of how multilingual models conduct transferring between languages.

Keywords: Intent recognition · Slot filling · Code-mixing · Code-switching · Multilingual models

1 Introduction

The usability of task-oriented dialog (ToD) assistants depends crucially on their ability to process users' utterances in many languages. At the core of a task-oriented dialog assistant is a natural language understanding (NLU) component, which parses an input utterances into a semantic frame by means of intent recognition and slot filling [40]. Intent recognition tools identify user needs (such as buy a flight ticket). Slot filling extracts the intent's arguments (such as departure city and time).

Common approaches to training multilingual task-oriented dialogue systems rely on (i) the abilities of pre-trained language models to transfer learning across languages and [5,25] and (ii) translate-and-align pipelines [18].

E. Burnaev et al. (Eds.): AIST 2021, CCIS 1573, pp. 3–16, 2022.
https://doi.org/10.1007/978-3-031-15168-2_1

The latter group of methods leverages upon pre-training on large amounts of raw textual data for many languages. Different learning objectives are used to train representations aligned across languages. The alignment can be further improved by means of task-specific label projection [27] and representation alignment [14] methods.

The former group of methods utilizes off-the-shelf machine translation engines to translate (i) the training data from resource-rich languages (almost exclusively English) to target languages or (ii) the evaluation data from target languages into English [18]. Further word alignment techniques help to match slot-level annotations [9, 30]. Finally, a monolingual model is trained to make desired predictions.

A number of datasets for cross-lingual NLU have been developed. To name a few, MultiAtis++ [44], covering seven languages across four language families, contains dialogues related to a single domain – air travels. MTOP [23] comprises six languages and 11 domains. xSID [13] is an evaluation-only small-scale dataset, collected for 12 languages and six domains.

Recent research has adopted a new experimental direction aimed at developing cross-lingual augmentation techniques, which learn the inter-lingual semantics across languages [9, 15, 21, 26, 31]. These work seek to simulate code-switching, a phenomenon where speakers use multiple languages mixed up [35]. Experimental results consistently show that augmentation with synthetic code-switched data leads to significantly improved performance for cross-lingual NLU tasks. What is more, leveraging upon these datasets in practice meets the needs of multicultural and multilingual communities. To the best of our knowledge, large-scale code-switching ToD corpora do not exist.

This paper extends the ongoing research on exploring benefits from synthetic code-switching to cross-lingual NLU. Our approach to generating code-switched utterances relies on grey-box adversarial attacks on the NLU model. We perturb the source utterances by replacing words or phrases with their translations to another language. Next, perturbed utterances are fed to the NLU model. Increases in the loss function indicate difficulties in making predictions for the utterance. This way, we can (i) generate code-switched adversarial utterances, (ii) discover insights on how code-switching with different languages impacts the performance of the target language, (iii) gather augmentation data for further fine-tuning of the language model. To sum, our contributions are[1]:

1. We implement several simple heuristics to generate code-switched utterances based on monolingual data from an NLU benchmark;
2. We show-case that monolingual models fail to process code-switched utterances. At the same time, cross-lingual models cope much better with such texts;
3. We show that fine-tuning of the language model on code-switched utterances improves the overall semantic parsing performance by up to a 2-fold increase.

---

[1] Our code can be found at https://github.com/PragmaticsLab/CodeSwitchingAdversarial.

2 Related Work

Generation of Code-Switched Text has been explored as a standalone task [17, 22, 33, 34, 39, 42] and as a way to augment training data for cross-lingual applications, including task-oriented dialog systems [9, 15, 21, 26, 31], machine translation [1, 12, 16], natural language inference and question answering [37], speech recognition [46].

Methods for generating code-switched text range from simplistic re-writing of some words in the target script [12] to adversarial attacks on cross-lingual pre-trained language models [39] and building complex hierarchical VAE-based (Variational AutoEncoders) models [34]. The vast majority of methods utilize machine translation engines [37], parallel datasets [1, 12, 16, 42] or bilingual lexicons [39] to replace the segment of the input text with its translations. Bilingual lexicons may be induced from the parallel corpus with the help of soft alignment, produced by attention mechanisms [22, 26]. Pointer networks can be used to select segments for further replacement [17, 42]. If natural code-switched data is available, such segments can be identified with a sequence labeling model [16]. Other methods rely on linguistic theories of code-switching. To this end, GCM toolkit [33] leverages two linguistic theories, which help to identify segments where code-switching may occur by aligning words and mapping parse trees of parallel sentences.

The quality of generated code-switched texts is evaluated by (i) **intrinsic** text properties, such as code-switching ratio, length distribution, and (ii) **extrinsic** measures, ranging from the perplexity of external languages model to the downstream task performance, in which code-switched data was used for augmentation [34].

Natural Language Understanding in the ToD domain has two main goals, namely intent recognition, and slot filling [32]. Intent recognition assigns a user utterance with one of the pre-defined intent labels. Thus, intent recognition is usually tackled with classification methods. Slot filling seeks to find arguments of the assigned intent and is modeled as a sequence modeling problem. For example, the utterance *I need a flight from Moscow to Tel Aviv on 2nd of December* should be assigned with the intent label find flight; three slots may be filled: departure city, arrival city, date. These two interdependent NLU tasks are frequently approached via multitask learning, in which a joint model is trained to recognize intents and fill in slots simultaneously [3, 20, 43].

Adversarial Attacks on natural language models has been categorized with respect to (i) what kind of information is provided from the model and (ii) what kind of perturbation is applied to the input text [29]. White-box attacks [8] have access to the whole model's inner workings. On the opposite side, black-box attacks [11] do not have any knowledge about the model. Grey-box attacks [45] access predicted probabilities and loss function values. Perturbations can be applied at char-, token-, and sentence-levels [4, 11, 24].

Other Related Research Directions include **code-switching detection** [28, 38], evaluation of pre-trained language models' **robustness** to code-switching [41], analysis of language model's inner workings with respect to code-switched inputs [36], **bench-marking** downstream tasks in code-switched data [2, 19].

3   Our Approach

In our work we train multilingual language models for the joint intent recognition and slot-filling task.

3.1   Dataset

We chose MultiAtis++ dataset [44] as the main source of data. This dataset contains seven languages from three language families - Indo-European (English, German, Portuguese, Spanish, and French), Japanese, and Sino-Tibetan (Chinese). The dataset is a parallel corpus for classifying intents and slot labeling - in 2020 it was translated from English to the other six languages. The training set contains 4978 samples for each language; the test set contains 893 samples per language. Each object in the dataset consists of a sentence, slot labels in BIO format, and the intent label.

3.2   Joint Intent Recognition and Slot-Filling

We train a single model for the joint intent recognition and slot-filling task. The model has two heads; the first one predicts intents, and the second one predicts slots. We had trained two different models as a backbone - m-BERT and XLM-RoBERTa.

We aim at comparing two setups: (i) training on the whole dataset and (ii) only on its English subset followed by zero-shot inference for other languages. For convenience, we propose short names for the four models trained during the research - **xlm-r, xlm-r-en, m-bert, m-bert-en**.

We measure our models' quality with three metrics: **intent accuracy**, **F1 score** for slots (we used micro-averaging by classes) and the proportion of sentences where we correctly classified everything - both intent and all slots - **semantic accuracy**.

3.3   Code-switching Generation

We propose two variants of gray-box adversarial attacks. During the attack, we have access to the model's loss of input data. We strive to create an attack so that the resulting adversarial perturbation of the source sentence is as close as possible to the realistic code-switching. Quality evaluation at such adversarial attacks can act as a lower bound for corresponding models' quality in similar problems in the presence of real code-switching in input data.

We focus mainly on the lexical aspect of code-switching when some words are replaced with their substitutes from other languages. We replace some tokens in the source sentence with their equivalents from the attacking languages during the attack. The method to determine the replacement depends on which exactly attack is used. Since most people who can use code-switching are bilinguals, in our work, we propose to analyze attacks consisting in embedding one language into another.

Algorithm 1. General view of the attack

Require: Sentence and label x, y; source model $\mathcal{M}$; embedded target language L
Ensure: Adversarial sample x'
   $\mathcal{L}_x$ = GetLoss($\mathcal{M}$, x, y)
   for i in permutation(len(x)) **do**
      Candidates = GetCandidates($\mathcal{M}$, x, y, token_id = i)
      Losses = GetLoss($\mathcal{M}$, Candidates)
      if Candidates and max(Losses) $\geq \mathcal{L}_x$ **then**
         $\mathcal{L}_x$ = max(Losses)
         x, y = Candidates[argmax(Losses)]
      end if
   end for
      return x

Overview of the Attacks. The general attack scheme (Algorithm 1) is the same for both proposed attacks. We offer the following attack pattern in our work: a source model, a pair of sample sentences and labels, and embedded target language. Then we iterate over the tokens in the sample sentence and strive to replace them with their equivalent from the embedded target language. If changing the token to its equivalent increases the source model's loss, we replace the token with the proposed candidate. The difference between the two methods consists in the way they generate replacement candidates.

Word-level Adversaries. The first attack (Algorithm 2) generates target embedded language substitutions by translating single tokens into the corresponding language. Attacking this way, we make a rough lower bound since we do not consider the context of the sentences and the ambiguity of words during the attack. To translate words into other languages, we use the large-scale many-to-many machine translation model M2M-100 from Facebook [10]. You can see an example of this attack in table 1.

Algorithm 2. Word-level attack

Require: Machine translation model T
   function GETCANDIDATES($\mathcal{M}$, x, y, token_id)
      if x[token_id] in $T[L]$ **then**
         tokens = $T[L]$[x[token_id]]
         x[token_id] = tokens
         y[token_id] = ExtendSlotLabels(y[token_id], len(tokens))
      end if
      return x, y
   end function

Table 1. Example of attacking XLM-RoBERTa (xlm-r) with word-level attack.

Utterance en	What are the flights from las vegas to Ontario
Utterance adv	What sind die flights from las vegas to Ontario

Phrase-level Adversaries. The second attack (Algorithm 3) generates equivalents from other languages by building alignments between sentences in different languages. One sentence is a translation of another; we utilize the fact that we have a parallel dataset. Candidates for each token are defined as tokens from the embedded sentence into which the token was aligned. For aligning sentences, we use the awesome-align model based on m-BERT [7]. You can see an example of this attack in table 2.

Algorithm 3. Phrase-level attack

Require: Sentences alignment A
  function GETCANDIDATES($\mathcal{M}$, x, y, token_id)
    if x[token_id] in $A[L]$ **then**
      tokens = $A[L]$[x[token_id]]
      x[token_id] = tokens
      y[token_id] = ExtendSlotLabels(y[token_id], len(tokens))
    end if
    return x, y
  end function

Table 2. Example of attacking XLM-RoBERTa (xlm-r) with phrase-level attack.

Utterance en	Please find flights available from kansas city to newark
Utterance adv	Encontre find flights disponíveis from kansas city para newark

3.4 Adversarial Pre-training Method Protects from Adversarial Attacks

Adversarial pre-training protects the model against the proposed adversarial attacks. It most likely allows the model to increase the performance not only at adversarial perturbations but also on real data with code-switching. However, this is only a hypothesis since there is no real-life code-switched ToD data.

    The adversarial pre-training method relies on domain adaptation techniques and has several steps:

1. Generating adversarial training set for masked language modeling task.
2. Fine-tuning language model's body on the new generated set in masked language modeling task.
3. Loading fine-tuned model's body before training for joint intent classification and slot labeling task.

Generating Adversarial Training Set. To generate an adversarial training set, we use an adaptation of the phrase-level algorithm of the adversarial attack (Algorithm 4). The difference is that tokens are replaced with their equivalents with the probability of 0.5. Thus, a trained model is not required to generate the sample. The adversarial training set is a concatenation of generated sets for all languages in the dataset except English. Each subset is generated by embedding the target language into the training set of the MultiAtis++ dataset in English. After generation, we get six subsets of 4884 sentences each. The final adversarial training set consists of 29304 sentences; we divide it into training and test sets in a ratio of 9 to 1.

Algorithm 4. Generating adversarial training set

Require: Training dataset X, set of embedded languages $L_1, \ldots L_n$
Ensure: Adversarial training set X'
  X' = [ ]
  for L in $L_1, \ldots L_n$ **do**
    for x in X **do**
      for i in permutation(len(x)) **do**
        Candidates = GetCandidates($\mathcal{M}$, x, y, token_id = i)
        if Candidates and $\mathcal{U}(0, 1)$ ¿ 0.5 **then**
          x, _ = random.choice(Candidates)
        end if
      end for
      X'.append(x)
    end for
  end for
  return X'

Fine-tuning Model's Body. After generating the adversarial training set, we fine-tune the pre-trained multilingual model. The model is trained with the masked language modeling objective [6]. We select 15% of tokens and predict them using the model to train a model for such a task. 80% of the selected tokens are replaced with the mask token, 10% are replaced with random words from the model's dictionary, the remaining 10% remain unchanged. After fine-tuning, we dump the body of the model for future use.

Loading Fine-Tuned Model's Body. Before training the multilingual model for the task of join intent classification and slot labeling, we load the fine-tuned body of the model. For models that have been pre-trained using the adversarial pre-training method, we will add the suffix **adv** to the name.

4 Experimental Results

We will compare models trained only on the English training set (zero-shot models) and the whole training set (full models). We will evaluate the quality by three metrics -

accuracy for intents, f1-score for slots and semantic accuracy. We found that zero-shot models have significantly worse quality than full ones, not only in languages other than English but even in English.

Joint Intent Classification and Slot Labelling. We have achieved strong performance for the problem of classifying intents and filling in slots. On the test sample, full models showed an average of 97% correct answers for intents, and zero-shot ones, on average, 85%. Full models showed 0.93 f1-score for slots, zero-shot 0.68. Full models showed 79% of completely correctly classified sentences and zero-shot ones - about 26%. This shows that zero-shot learning is not capable of competing with full learning in this particular task. In the Fig 1, you can see the comparison between models and languages by Intent accuracy metric. The additional results are provided in the project's repository.

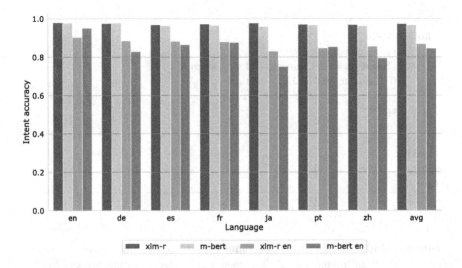

Fig. 1. Model comparison by **Intent accuracy** metric.

Attacking Models. We attacked all the models using our two algorithms. We found that the word-level attack turned out to be more advanced, leading to lower performance. For full models, the quality of intents fell from 98% to 88%, for zero-shot models from 92% to 77%. For full models, the quality by slots fell from 0.95 to 0.6, for zero-shot models from 0.88 to 0.48. For full models, the proportion of entirely correctly classified sentences fell from 83% to 14%, for zero-shot ones from 60% to 5%. Figure 2 compares results before and after the word-level attack with the Intent accuracy metric.

We also got that the phrase-level attack turned out to be softer and gave a higher quality compared to the word-level attack. For full models, the quality of intents fell from 98% to 95%, for zero-shot models from 92% to 80%. For full models, the quality by slots fell from 0.95 to 0.7, for zero-shot models from 0.88 to 0.55. For full models, the proportion of entirely correctly classified sentences fell from 83% to 35%, for zero-shot ones from 60% to 10%. In the Fig 3, you can see the quality comparison after the phrase-level attack for the Intent accuracy metric.

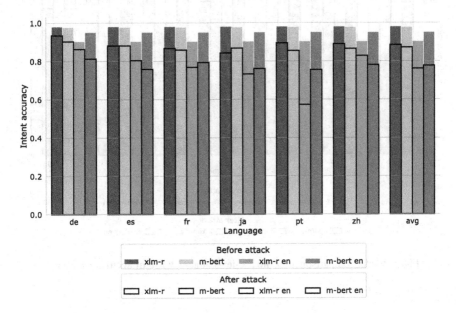

Fig. 2. Model comparison after **word-level** attack by **Intent accuracy** metric.

Adversarial Pre-training. To protect the models from attacks, we fine-tuned the bodies for both models and loaded them before training for the task of joint intent classification and slot filling. We found that the defense had almost no effect on the full models in terms of quality on the test set. For zero-shot models, the effect on the test set is ambiguous - the quality by intents fell for Asian languages but increased slightly for all others. As for the slots, we observe a negative effect for the m-BERT model and a positive effect for the XLM-RoBERTa model.

For the word-level attack, a slight deterioration in the quality of intents for Asian languages is noticeable, and a positive effect for other languages. After the adversarial pre-training, the quality of slots increased for all models, which ultimately results in an almost two-fold increase in the proportion of entirely correctly classified sentences for zero-shot models and about 15% relative improvement for full models.

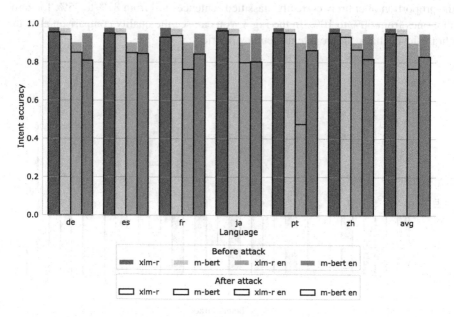

Fig. 3. Model comparison after **phrase-level** attack by **Intent accuracy** metric.

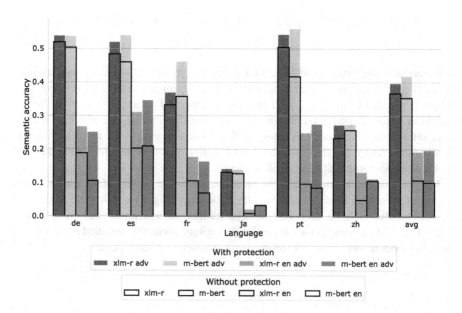

Fig. 4. Model comparison **with protection** after **phrase-level** attack by **Semantic accuracy** metric.

Again, for the phrase-level attack, a slight deterioration in the quality of intents for Asian languages is noticeable, and a positive effect for other languages. After the defense, the quality of slots dropped slightly for Asian languages and increased significantly for the rest. This results in a two-fold increase in the proportion of entirely correctly classified sentences for zero-shot models and about 15% relative improvement for full models (Fig 4).

5   Discussion

We approached the problem of recognizing intents and filling in slots for a multi-lingual ToD system. We study the effect of switching codes on two multi-lingual language models, XLM-RoBERTa and m-BERT. We showed that switching codes could become a noticeable problem when applying language models in practice using two gray-box attacks. However, the defense method shows promising results and helps to improve the quality after the model is attacked.

6   Conclusion

This paper presents an adversarial attack on multilingual task-oriented dialog (ToD) systems that simulates code-switching. This work is motivated by research and practical needs. First, the proposed attack reveals that pre-trained language models are vulnerable to synthetic code-switching. To this end, we develop a simplistic defense technique against code-switched adversaries. Second, our work is motivated by the practical needs of multilingual ToDs to cope with code-switching, which is seen as an essential phenomenon in multicultural societies. Future work directions include evaluating how plausible and naturally-sounding code-switched adversaries are and adopting similar approaches to model-independent black-box scenarios.

Acknowledgements. The article was prepared within the framework of the HSE University Basic Research Program.

References

1. Abdul-Mageed, M., Lakshmanan, L.V.: Exploring text-to-text transformers for English to Hinglish machine translation with synthetic code-mixing. In: Proceedings of the 2021 Conference of the North American Chapter of the Association for Computational Linguistics: Human Language Technologies, pp. 36–46 (2021)
2. Aguilar, G., Kar, S., Solorio, T.: LinCE: A centralized benchmark for linguistic code-switching evaluation. In: Proceedings of The 12th Language Resources and Evaluation Conference. pp. 1803–1813. European Language Resources Association, Marseille, France (2020). https://www.aclweb.org/anthology/2020.lrec-1.223
3. Chen, Q., Zhuo, Z., Wang, W.: BERT for joint intent classification and slot filling (2019)
4. Cheng, M., Yi, J., Chen, P.Y., Zhang, H., Hsieh, C.J.: Seq2sick: Evaluating the robustness of sequence-to-sequence models with adversarial examples. In: Proceedings of the AAAI Conference on Artificial Intelligence, vol. 34, pp. 3601–3608 (2020)

5. Conneau, A.: Unsupervised cross-lingual representation learning at scale(2019). arXiv preprint arXiv:1911.02116
6. Devlin, J., Chang, M.W., Lee, K., Toutanova, K.: BERT: Pre-training of deep bidirectional transformers for language understanding. In: Proceedings of the 2019 Conference of the North American Chapter of the Association for Computational Linguistics: Human Language Technologies, vol. 1 (Long and Short Papers), pp. 4171–4186 (2019)
7. Dou, Z.Y., Neubig, G.: Word alignment by fine-tuning embeddings on parallel corpora. In: Conference of the European Chapter of the Association for Computational Linguistics (EACL) (2021)
8. Ebrahimi, J., Rao, A., Lowd, D., Dou, D.: Hotflip: White-box adversarial examples for text classification. In: Proceedings of the 56th Annual Meeting of the Association for Computational Linguistics, vol. 2, pp. 31–36 (2018)
9. Einolghozati, A., Arora, A., Lecanda, L.S.M., Kumar, A., Gupta, S.: El volumen louder por favor: code-switching in task-oriented semantic parsing. In: Proceedings of the 16th Conference of the European Chapter of the Association for Computational Linguistics: Main Volume, pp. 1009–1021 (2021)
10. Fan, A., et al.: Beyond English-centric multilingual machine translation (2020). arXiv preprint
11. Gao, J., Lanchantin, J., Soffa, M.L., Qi, Y.: Black-box generation of adversarial text sequences to evade deep learning classifiers. In: 2018 IEEE Security and Privacy Workshops (SPW), pp. 50–56, IEEE (2018)
12. Gautam, D., Kodali, P., Gupta, K., Goel, A., Shrivastava, M., Kumaraguru, P.: Comet: Towards code-mixed translation using parallel monolingual sentences. In: Proceedings of the Fifth Workshop on Computational Approaches to Linguistic Code-Switching, pp. 47–55 (2021)
13. van der Goot, R., et al.: From masked language modeling to translation: Non-english auxiliary tasks improve zero-shot spoken language understanding (2021)
14. Gritta, M., Iacobacci, I.: Xeroalign: Zero-shot cross-lingual transformer alignment (2021). arXiv preprint arXiv:2105.02472
15. Guo, Y.: Learning from multiple noisy augmented data sets for better cross-lingual spoken language understanding (2021). arXiv preprint arXiv:2109.01583
16. Gupta, A., Vavre, A., Sarawagi, S.: Training data augmentation for code-mixed translation. In: Proceedings of the 2021 Conference of the North American Chapter of the Association for Computational Linguistics: Human Language Technologies, pp. 5760–5766 (2021)
17. Gupta, D., Ekbal, A., Bhattacharyya, P.: A semi-supervised approach to generate the code-mixed text using pre-trained encoder and transfer learning. In: Proceedings of the 2020 Conference on Empirical Methods in Natural Language Processing: Findings, pp. 2267–2280 (2020)
18. Hu, J., Ruder, S., Siddhant, A., Neubig, G., Firat, O., Johnson, M.: Xtreme: A massively multilingual multi-task benchmark for evaluating cross-lingual generalisation. In: International Conference on Machine Learning, PMLR, pp. 4411–4421 (2020)
19. Khanuja, S., Dandapat, S., Srinivasan, A., Sitaram, S., Choudhury, M.: Gluecos: An evaluation benchmark for code-switched NLP. In: Proceedings of the 58th Annual Meeting of the Association for Computational Linguistics, pp. 3575–3585 (2020)
20. Kim, Y., Kim, D., Kumar, A., Sarikaya, R.: Efficient large-scale neural domain classification with personalized attention. In: Gurevych, I., Miyao, Y. (eds.) Proceedings of the 56th Annual Meeting of the Association for Computational Linguistics, ACL 2018, Melbourne, Australia, July 15–20, vol. 1, pp. 2214–2224, Association for Computational Linguistics (2018). 10.18653/v1/P18-1206, https://aclanthology.org/P18-1206/

21. Krishnan, J., Anastasopoulos, A., Purohit, H., Rangwala, H.: Multilingual code-switching for zero-shot cross-lingual intent prediction and slot filling (2021). arXiv preprint arXiv:2103.07792
22. Lee, G., Li, H.: Modeling code-switch languages using bilingual parallel corpus. In: Proceedings of the 58th Annual Meeting of the Association for Computational Linguistics. pp. 860–870 (2020)
23. Li, H., Arora, A., Chen, S., Gupta, A., Gupta, S., Mehdad, Y.: MTOP: a comprehensive multilingual task-oriented semantic parsing benchmark. In: Proceedings of the 16th Conference of the European Chapter of the Association for Computational Linguistics: Main Volume. pp. 2950–2962 (2021)
24. Li, L., Ma, R., Guo, Q., Xue, X., Qiu, X.: BERT-attack: Adversarial attack against BERT using BERT. In: Proceedings of the 2020 Conference on Empirical Methods in Natural Language Processing (EMNLP), pp. 6193–6202 (2020)
25. Liu, Y., et al.: Multilingual denoising pre-training for neural machine translation. Trans. Assoc. Comput. Linguis. **8**, 726–742 (2020)
26. Liu, Z., Winata, G.I., Lin, Z., Xu, P., Fung, P.N.: Attention-informed mixed-language training for zero-shot cross-lingual task-oriented dialogue systems. In: Proceedings of the AAAI Conference on Artificial Intelligence (2020)
27. Liu, Z., Winata, G.I., Xu, P., Lin, Z., Fung, P.: Cross-lingual spoken language understanding with regularized representation alignment. In: Proceedings of the 2020 Conference on Empirical Methods in Natural Language Processing (EMNLP), pp. 7241–7251 (2020)
28. Mave, D., Maharjan, S., Solorio, T.: Language identification and analysis of code-switched social media text. In: Proceedings of the Third Workshop on Computational Approaches to Linguistic Code-Switching. pp. 51–61, Association for Computational Linguistics, Melbourne, Australia (2018). 10.18653/v1/W18-3206, https://aclanthology.org/W18-3206
29. Morris, J., Lifland, E., Yoo, J.Y., Grigsby, J., Jin, D., Qi, Y.: Textattack: a framework for adversarial attacks, data augmentation, and adversarial training in NLP. In: Proceedings of the 2020 Conference on Empirical Methods in Natural Language Processing: System Demonstrations, pp. 119–126 (2020)
30. Nicosia, M., Qu, Z., Altun, Y.: Translate & fill: Improving zero-shot multilingual semantic parsing with synthetic data (2021). arXiv preprint arXiv:2109.04319
31. Qin, L., Ni, M., Zhang, Y., Che, W.: CoSDA-ML: Multi-lingual code-switching data augmentation for zero-shot cross-lingual NLP (2020). arXiv preprint arXiv:2006.06402
32. Razumovskaia, E., Glavaš, G., Majewska, O., Ponti, E.M., Korhonen, A., Vulić, I.: Crossing the conversational chasm: A primer on natural language processing for multilingual task-oriented dialogue systems (2021)
33. Rizvi, M.S.Z., Srinivasan, A., Ganu, T., Choudhury, M., Sitaram, S.: GCM: A toolkit for generating synthetic code-mixed text. In: Proceedings of the 16th Conference of the European Chapter of the Association for Computational Linguistics: System Demonstrations, pp. 205–211 (2021)
34. Samanta, B., Reddy, S., Jagirdar, H., Ganguly, N., Chakrabarti, S.: A deep generative model for code-switched text (2019). arXiv preprint arXiv:1906.08972
35. Sankoff, D., Poplack, S.: A formal grammar for code-switching. Res. Lang. & Soc. Interac. **14**(1), 3–45 (1981)
36. Santy, S., Srinivasan, A., Choudhury, M.: BERTologicomix: How does code-mixing interact with multilingual BERT? In: Proceedings of the Second Workshop on Domain Adaptation for NLP, pp. 111–121 (2021)
37. Singh, J., McCann, B., Keskar, N.S., Xiong, C., Socher, R.: XLDA: Cross-lingual data augmentation for natural language inference and question answering (2019). arXiv preprint arXiv:1905.11471

38. Sravani, D., Kameswari, L., Mamidi, R.: Political discourse analysis: a case study of code mixing and code switching in political speeches. In: Proceedings of the Fifth Workshop on Computational Approaches to Linguistic Code-Switching. pp. 1–5, Association for Computational Linguistics (2021). 10.18653/v1/2021.calcs-1.1, https://aclanthology.org/2021.calcs-1.1

39. Tan, S., Joty, S.: Code-mixing on sesame street: dawn of the adversarial polyglots. In: Proceedings of the 2021 Conference of the North American Chapter of the Association for Computational Linguistics: Human Language Technologies, pp. 3596–3616 (2021)

40. Tur, G., Hakkani-Tür, D., Heck, L.: What is left to be understood in ATIS? In: 2010 IEEE Spoken Language Technology Workshop, IEEE, pp, 19–24 (2010)

41. Winata, G.I., Cahyawijaya, S., Liu, Z., Lin, Z., Madotto, A., Fung, P.: Are multilingual models effective in code-switching? NAACL **2021**, 142 (2021)

42. Winata, G.I., Madotto, A., Wu, C.S., Fung, P.: Code-switched language models using neural based synthetic data from parallel sentences. In: Proceedings of the 23rd Conference on Computational Natural Language Learning (CoNLL), pp. 271–280 (2019)

43. Wu, C.S., Hoi, S.C., Socher, R., Xiong, C.: TOD-BERT: Pre-trained natural language understanding for task-oriented dialogue. In: Proceedings of the 2020 Conference on Empirical Methods in Natural Language Processing (EMNLP). pp. 917–929. Association for Computational Linguistics (2020). 10.18653/v1/2020.emnlp-main.66, https://aclanthology.org/2020.emnlp-main.66

44. Xu, W., Haider, B., Mansour, S.: End-to-end slot alignment and recognition for cross-lingual NLU. In: Proceedings of the 2020 Conference on Empirical Methods in Natural Language Processing (EMNLP), pp. 5052–5063 (2020)

45. Xu, Y., Zhong, X., Yepes, A.J., Lau, J.H.: Grey-box adversarial attack and defence for sentiment classification. In: Proceedings of the 2021 Conference of the North American Chapter of the Association for Computational Linguistics: Human Language Technologies, pp. 4078–4087 (2021)

46. Yılmaz, E., Heuvel, H.v.d., van Leeuwen, D.A.: Acoustic and textual data augmentation for improved asr of code-switching speech (2018). arXiv preprint arXiv:1807.10945

Extracting Software Requirements
from Unstructured Documents

Vladimir Ivanov[1(✉)], Andrey Sadovykh[1,2], Alexandr Naumchev[1],
Alessandra Bagnato[2], and Kirill Yakovlev[1]

[1] Innopolis University, Innopolis, Russia
`v.ivanov@innopolis.ru`
[2] Softeam, Paris, France
`http://university.innopolis.ru, http://www.softeam.com`

Abstract. Requirements identification in textual documents or extraction is a tedious and error prone task that many researchers suggest automating. We manually annotated the PURE dataset and thus created a new one containing both requirements and non-requirements. Using this dataset, we fine-tuned the BERT model and compare the results with several baselines such as fastText and ELMo. In order to evaluate the model on semantically more complex documents we compare the PURE dataset results with experiments on Request For Information (RFI) documents. The RFIs often include software requirements, but in a less standardized way. The fine-tuned BERT showed promising results on PURE dataset on the binary sentence classification task. Comparing with previous and recent studies dealing with constrained inputs, our approach demonstrates high performance in terms of precision and recall metrics, while being agnostic to the unstructured textual input.

Keywords: Software requirements · Requirements elicitation ·
BERT · FastText · ELMo · Sentence classification

1 Introduction

Identifying and understanding technical requirements is a challenging task for many reasons. Requirements come from various sources and in various forms. These different forms may serve different purposes and levels, from initial project proposals down to detailed contracts of individual methods. The forms that appear earlier in the software process are consumed by more business-oriented stakeholders. Technical people, such as testers, require detailed specifications produced later in the software process to convert them into tests. The evolving form of describing the initial problem is natural and reflects the process of understanding the problem better with time. The available requirement datasets reflect those variations.

One of such is the PURE dataset [1] presented by A. Ferrari et al., which includes 79 Software Requirements Specification (SRS) documents. Mainly this

E. Burnaev et al. (Eds.): AIST 2021, CCIS 1573, pp. 17–29, 2022.
https://doi.org/10.1007/978-3-031-15168-2_2

set is based on the Web search and includes documents with different peculiarities as well as a lexicon with the widespread writing style of requirements. Authors argue that this dataset fairly fits for various Natural Language Processing (NLP) tasks such as requirements categorisation, ambiguity detection, and equivalent requirements identification, etc. But at the same time, additional expertise, as well as an annotation process, are primarily required. However, some other datasets might be challenging. The Request for Information (RFI) documents in itself specify an initial set of "best wishes" for a required solution as well as a variety of constraints in terms of deployment, compatibility or guarantees for maintenance and support. Those documents are meant for engineers though the language used, format, styles and structure are far from being standard. Those parameters largely depend on the organization that issued those documents, personnel involved in designing specifications and traditional practices of the domain the organization belongs to. On the other side, the provider of the solution must quickly qualify the RFI and prepare a well-thought offer that would maximize the value for the customer at minimum costs for the provider. Such analysis is mostly done manually that is costly, time-consuming and error-prone.

In this paper, we present an approach for extraction of requirements coming from unstructured text sources. We fine-tuned the Bidirectional Encoder Representations from Transformers (BERT) [2] model with PURE-based manually annotated dataset of 7,745 sentences to identify technical requirements in natural text. The ReqExp showed 92% precision and 80% recall on the PURE dataset. In order to evaluate BERT, we compared the results with previous state-or-art methods such as fastText and Embeddings from Language Model (ELMo) with an SVM classifier on the same PURE-based dataset, where BERT showed superior results in the precision, but worse results in the recall. Afterwards we carried-out additional experiments with more semantically difficult texts from RFIs. We applied all the candidate models trained on PURE-based dataset to a private dataset of 380 sentences extracted from RFI documents in order to evaluate the possibility to transfer the result to a new domain. BERT again showed superior results in precision (90%), but inferior in recall (71%). With a superior recall metrics fastText (82%) and ELMo (72%) baseline methods showed less false negatives in detecting requirements, i.e. passed through less of undetected requirements. The results obtained in the limited set of experiments showed optimistic outcomes. We ran an additional experiment with a larger training set, consisted of all the text samples from the PURE and obtained precision of 86% and recall of 84% that suggests further fine-tuning of ReqExp on a larger dataset. Moreover, the analysis of related work indicated that the ReqExp approach has certain advantages, since, for example, it is agnostic to the type and the structure of input documents.

Unfortunately, the datasets of comparable studies are not available for a thorough analysis. In order to avoid the same problem with our study, we provide our dataset with manually annotated sentences from the PURE corpus to the community for comparison studies. The initial intention for this study came from to Softeam company from France. They kindly provided the dataset of RFI documents and endorsed our work. Softeam colleagues confirmed that the results are promising for the industry with a high potential to be used in the company's products and processes.

To summarize this paper, the contributions of this work are: (1) adapted PURE-based dataset for classification tasks with manually annotated sentences containing both requirements and non-requirements; (2) the description of BERT-based tool chain applying NLP for requirements extraction; (3) the results of evaluation on SRS sentences from the PURE dataset in comparison with baseline methods – fastText and ELMo; (4) the results of evaluation on a provide dataset of RFI documents; (5) analysis of comparable studies. Therefore, the current paper starts with the outline of the ReqExp approach in Sect. 2 describing the conceptual method and the dataset construction for profiling BERT neural network, followed by classifier models details in Sect. 3 , continued by validation of the prototype on PURE-based dataset and real RFI documents in Sect. 4, to be concluded by a comparison with the related work in Sect. 5.

2   Methods and Datasets

In this section we describe the process of building a binary sentence classifier. This process has several steps presented in Fig. 1. The ultimate goal was to efficiently build a training dataset that can be used to fine-tune the BERT model and compare it with other advanced NLP methods. Bidirectional Encoder Representations from Transformers (BERT) is a model by Google pre-trained on a huge text corpus [2]. BERT was tested on multiple NLP tasks and achieved state-of-the-art performance. A common way to use BERT for a domain-specific task, such as software requirements extraction, is to fine-tune the model i.e., to optimize model's weights on a domain-specific dataset. To solve a binary classification problem, such dataset should contain two subsets of sentences: (i) a subset of sentences that contain requirements, and (ii) a subset of sentences that does not contain requirements. However, there were no such dataset readily available for requirements extraction.

Thus, the first step of our approach is creating the domain-specific training dataset for fine-tuning. To speed up the process of corpus creation, we decided to process an existing corpus of software specification documents. In [1], Ferrari et al. presented a text corpus of public requirements documents (PURE)[1] The dataset contains 79 publicly available natural language requirements documents. We applied PURE corpus as our main source of both requirements and non-requirements for designing our own dataset. Initially, documents were presented in the form of raw text with different format, structure and writing style.

At the beginning we made an attempt to extract sentences with requirements using some parsing and extraction engines. Unfortunately, this idea was rejected due to unsatisfactory quality of extracted text samples. Despite that authors did provide a set of text requirements in the form of *.json* files to simplify the

---

[1] Data of the corpus can be found here: https://zenodo.org/record/1414117.

extraction process, this was not enough to design a representative subset for each class. It was decided to process documents manually preserving initial syntax and semantics in the extracted text samples. Still this process was associated with some challenges, which required us to apply the following conventions:

– Due to different length and structure of requirements, it was decided to assume one sentence as one data unit. It means that a requirement, which consists of several sentences, was divided into several units.
– Sometimes requirements were written in the form of a list with many elements. In such case, this list was transformed into one sentence separating parts by commas. In some rare cases, where elements were individual sentences, such requirement was divided into several sentences accordingly.

Overall, we extracted 7,745 sentences, where 4,145 were requirements and 3,600 were non-requirements from 30 documents. The extraction was done from the appropriate sections of the documents. Usually each document provides structural elements like table of elements, requirements-focused sections with appropriate names or contextual footnotes. The latter one might have some specific numbering used in a document and initially explained by the authors, or lexical elements that are usually inherent for requirements such as modal verbs like 'must', 'shall', 'should', etc. When identifying requirements and non-requirements we followed principle described in [3].

Sentences with requirements were manually extracted by one specialist in Software Engineering area and then independently validated by another expert from our research group. Some special cases were separately discussed in the group to make annotation more consistent. In the case when there was no agreement concerning which label should be utilized to those text samples, they were removed to maintain consistency of the set.

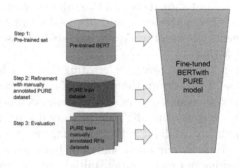

Fig. 1. The training process to obtain a model for binary sentence classification for requirements extraction.

3 Models for Sentence Classification

For this study we chose two baseline approaches, which preceded the BERT architecture, for binary sentence classification. Specifically we considered fast-Text model and ELMo embeddings with a classical machine learning classifier. The latter one was chosen through an experiment stage to choose the most promising model based on the most explicit and informative metrics during comparison.

3.1 FastText Baseline Model

fastText is a method for constructing word embeddings based on the Word2Vec principle, but with some extension. An embedding for each word is learnt from embedding vectors for character n-grams that constitute the word. Thus, each word vector is represented as the sum of the n-gram vectors. This extends the Word2Vec type models with subword information. Once a word is represented using character n-grams, a skipgram model is trained to learn the embeddings. This approach provides a fast and scalable solution, allowing to train models on large corpora [4]. For the down-stream task (binary classification of sentences) we use a linear classifier that is default (a multinomial logistic regression) in the fastText library.

An important part in designing predictive models is associated with finding the best hyper-parameters. Usually, it is a cumbersome and time-consuming task, especially when doing it manually. For this purpose, we applied an auto-tune feature located inside the fastText functionality by using the validation set presented earlier[2].

3.2 ELMo with SVM Classifier

Embeddings from Language Models, or ELMo is an advanced method for constructing a contextualized word representation. The core ability is to model different characteristics of word use together with linguistic context. Specifically, these representations are internal states of a deep bidirectional language model that is pre-trained on a large corpus. ELMo maximizes the log-likelihood of the forward and backward directions in a sequence of words [5]. These produced vectors can be applied as features for various NLP purposes.

To make a decision for distinguishing among sentences with requirement and without, one needs to use a classifier on top of the features extracted from text. To find the best classifier, we compared several popular machine learning models such as Logistic Regression, Random Forest classifier, Support Vector Machines and Multi-Layer Perceptron. To this end the Grid Search was applied with specific set of initially predefined hyperparameters for each model. The process of assessment was based entirely on the training set. As a result we chose with Support Vector Machines with polynomial kernel that showed the

---

[2] https://fasttext.cc/docs/en/autotune.html.

highest result in terms of F1-score. The results related to this baseline will be marked as "ELMo+SVM" in the following sections.

3.3 BERT-Based Approach

Recently using a large pre-trained language model has shown a breakthrough in many tasks of natural language processing including text classification. The baseline approach to transfer learning typically implies pre-training a neural network on a large corpus of texts with further fine-tuning the model on a specific task. In this study, we use pre-trained BERT [2] model[3] as a baseline. In all experiments with BERT, we adopted the implementation of fine-tuning process for binary text classification presented in [6]. Hyper-parameters used in fine-tuning are the following: batch size is 64, learning rate is $2 \cdot 10^{-5}$, max sequence length is 100, number of epochs is 2. Overall schema is presented in Fig. 1.

4   Experimental Setup and Validation

4.1   Experimental Setup

Empirical evaluation of model performance was carried out in two phases. First, we used a classical approach with splitting dataset into disjoint sets of sentences (namely, 'train', 'test' and 'validation' sets). All models were trained on the same 'train' subset and tested on the same 'test' part from the PURE dataset. We did not apply cross-validation due to the special rule of splitting collection into train and test sets, i.e. *sentences from some document should appear in either the train set or in the test set*. The dataset as well as the split is available for the sake of reproducibility and comparison of results.

Second, we used an out-of-sample data for evaluation of the classifiers. To this end we tested the performance on sentences manually extracted from the RFI documents. All models were trained on the same 'train' part and tested on the same group of sentences extracted from RFI. We argue that such experimental conditions are more relevant in practice. Indeed, we expected decreasing of the performance on the RFI documents. This shows that fine-tuning of large pre-trained language models is not a panacea in practice, however it gives a viable alternative to rule-based tools. Finally, we show that adding more data on the fine-tuning phase leads to better performance on the out-of-sample data.

4.2   Train, Test and Validation Sets

As for the data split, we had 7,745 sentences extracted from the PURE dataset to split them up into train, test and validation subsets by the parts of approximately 70%, 20 % and 10% accordingly. In order to consistently separate our

---

[3] Specifically, we use the BERT-base uncased model from https://github.com/google-research/bert.

dataset, sentences from every document were fully added only to one specific subset. We found a particular combination of documents for each subset that preserve the stated separation and at the same time preserves the balance of class labels. Statistics of the split is in sentences is the following: train set (5306); test set (1534); validation set (905). The dataset is available at https://zenodo.org/record/4630687.

4.3   Experiments with PURE Documents

The main purpose of this study was about assessing three major classification metrics: Precision, Recall, and F1-score. Those metrics are perceived as the golden standard in Machine Learning and Deep Learning areas.

Result of the first phase of evaluation is presented in Table 1. As it was expected, more advanced model (BERT) showed better results in terms of F1-score. BERT-based model showed high precision (0.92) and lower recall (0.8). Results of the BERT model is available at https://bit.ly/3oPElMm. However, the values of precision and recall metrics behave differently for the fastText and ELMo-based baselines. fastText-based classifier showed better Recall (0.93) comparing with other architectures. This property might be useful in some cases when it is necessary to extract more relevant sentences and text patterns associated with requirements.

Table 1. Results of experiments with PURE

Model	F1	P	R	TP	TN	FP	FN
fastText	.81	.72	**.93**	763	419	295	57
ELMo+SVM	.83	.78	.88	827	364	231	112
BERT	**.86**	.92	.80	841	407	69	217

4.4   Experiments with RFI Documents

In order to further evaluate our approach on less standardized requirements documents, we applied our prototypes to five anonymized documents provided by Softeam, France, as issued by their customers. We parsed the obtained documents to extract the paragraphs and then triggered the automatic requirements extraction from those paragraphs. After that we manually annotated all the sentences and thus assessed the four major metrics - the number of the true positives, the true negatives, the false positives and the false negatives. Then the other metrics such as precision, recall and F1-score were derived. The results can be found in Table 2.

To the validity of the results, it should be noted that the annotation was done by one single specialist who was aware of the context of the Request for Information (RFI) documents. This choice is justified to ensure the uniformity of the manual annotation, since we noticed that the understanding of whether a sentence is a requirement depends on context very much. In addition, it should be mentioned that our attempts of manual annotation by junior researchers were unfruitful since the results were completely unsatisfactory.

Table 2. Results of experiments with RFI

Model	F1	P	R	TP	TN	FP	FN
fastText	.65	.54	**.82**	146	83	120	31
ELMo+SVM	.69	.66	.73	177	48	89	66
BERT	**.80**	.90	.71	190	93	21	76

We also faced several challenges when were manually annotating the sentences from RFIs. First, RFI documents often specify technical requirements in the form of a question: "Will your solution provide a particular feature?" These questions are difficult to qualify as technical requirements in traditional sense. Second, a large part of RFI documents concern the constraints for the submission of the response such as delay, length, and language of the response. Should that be qualified as a requirement to documentation? All these issues are postponed to the follow up studies where we intend to refine the results. In particular, the study triggered more reflection about the nature of the requirements, since the specification practice is far from the form specified in the traditional requirements engineering standards.

Finally, we ran an additional experiment when we fine-tuned the best model on the sentences from train, validation and test sets altogether with further evaluation on the sentences from RFI documents. All parameters of fine-tuning were the same as in the first set of experiments. The performance has grown significantly (P = 0.86, R = 0.84, F1 = 0.85) compared to the initial setup with the fine-tuning on the train set only (Table 2). Thus, this indicates a possibility that the quality of the result can be improved further by improving and augmenting train dataset.

5   Related Work

The related work analysis in the present section relies on a recent mapping study of natural language processing (NLP) techniques in requirements engineering [7]. The research group analyzed 404 primary studies relevant to the NLP for requirements engineering (NLP4RE) domain. They mapped these studies into several dimensions and identified the key categories along each dimension. Here, we focus on the *Research Facet* that enumerates categories of research and of evaluation methods. Our contribution falls into the *solution proposal* category: we develop a new solution (a tool) for identifying likely requirements in software-related texts. We used the following methods for evaluating our proposed solution:

- *Experience Report*: "the result has been used on real examples, but not in the form of case studies or controlled experiments, the evidence of its use is collected informally or formally."
- *Laboratory Experiment with Software Subjects (LESS)*: "a laboratory experiment to compare the performance of newly proposed system with other existing systems."

Below we list studies from [7] that fall into the same categories as the ReqExp approach does. Some of the studies make strong assumptions about the respective input and/or output artifacts. We do not compare ReqExp with these studies because we do not make such assumptions: ReqExp takes as input arbitrary technical documents and identifies in them statements that look like potential system requirements. The approaches that work with documents from later stages of the requirements process are:

- ARSENAL [8] deals with detailed behavior specifications taking the form of "shall" statements and automatically converts them into formal models.
- AutoAnnotator [9] assumes to receive requirement documents as input and assigns semantic tags to these documents.
- CHOReOS [10] works with well-formed requirements specified by domain experts.
- Doc2Spec [11] works with JavaDoc and other in-code documentation.
- ELICA [12] extracts and classifies requirements-relevant information based on domain repositories and either existing requirement documents or elicitation transcripts.
- FENL [13] extracts software features from online software reviews.
- GUEST [14] works with well-formed requirement documents to extract goal and use case models.
- GaiusT [15] applies to regulatory documents and extracts statements of specific types with predefined structures.
- NLP-KAOS [16] analyzes research abstracts and produces KAOS goal diagrams from them.
- SAFE [17] analyzes descriptions and user reviews of mobile apps in application stores and extracts features from them.
- SNACC [18] automatically checks design information for compliance with regulatory documents.
- Text2Policy [19] processes requirements in natural language to search statements of a predefined type (access control policies).
- UCTD [20] takes use cases as input and produces use cases enriched with transaction trees.
- Guidance Tool [21] works with texts that already look much like use cases and transforms them into properly structured use cases.
- Text2UseCase [22] works with texts that already describe scenarios and transforms them into well-formed use cases.

The recent study [3] has explored the very related topic of demarcating requirements in textual specifications with ML-based approaches. The authors applied similar methods on a private dataset of 30 specification from industrial companies with presumably 2,145 sentences containing 663 requirements and 1,482 non-requirements candidates. They obtained comparable average precision of 81.2% and an average recall of 95.7%. Unfortunately, it impossible to compare our results on the same dataset. However, we should underline the major differences such as (1) our publicly available dataset is much larger and more balanced; (2) we trained and validated the model on completely different documents; (3) we applied much simpler end-to-end training method without any costly feature engineering; (4) that way our method is arguably more suitable for documents with complex semantics such as RFIs.

Another similar study [23] has presented an application of BERT for multiclass classification of requirements. Authors were focused on the following subtasks:

- Binary classification of functional (F) and non-functional (NFR) requirements on the original dataset.
- Binary and multiclass classification of the four most frequent NFR subclasses on all NFR in the original dataset.
- Multiclass classification of all NFR subclasses on NFR in the original NFR dataset.
- Binary classification of requirements based on functional and quality aspects using the relabeled NFR dataset by another study.

For each subtask they provided classification results in terms of precision, recall and F1-score. As it was expected, BERT showed the most promising results comparing with other baseline approaches.

6   Conclusions

Addressing the customers' needs involves a huge number of documents expressing requirements in various forms. In the initial steps of those interactions with customer the requirements are provided in RFI documents. Those documents often combine technical and commercial requirements as well as provide other constraints. The formats and styles of RFI vary a lot, which makes it difficult to technically qualify the needs and timely prepare an offer.

In this paper we propose a requirements extraction approach based on Machine Learning. We used BERT as the basis of the ML model, which we finetuned with a manually annotated dataset constructed from an existing requirements specification repository. We have compared the model to two competitive baselines (fastText-based and ELMo-based). Indeed, the study is preliminary as we just compare baselines models and do not propose any specific improvements for the architectures. However, the results of the baselines are quite promising and can be used in industry.

We validated our approach on five documents coming from Softeam company. The approach and validation results were compared to our previous studies. The five documents were annotated automatically at the sentence level (in total, 380 sentences with 6,676 tokens were extracted) and then evaluated by an expert to calculate the performance scores. This analysis showed that our approach superior, since it supports unconstrained textual specifications - it is agnostic to the format of the input textual document. The performance characteristics of our approach are on the comparable level even without costly feature engineering though a progress can be made. We may further evolve the prototype if, for example, we continue the fine-tuning. In addition, by combining train, test and validation sets into one training set has shown a dramatic increase of BERT performance on more semantically complex RFI documents ($P = 0.86$, $R = 0.84$, $F1 = 0.85$). This suggests usefulness of further work on manual annotation of the remaining documents from the PURE dataset and augmenting the train set.

The collaboration with Softeam made us realize the industrial usefulness of the approach in real cases as it can improve current tools and service by that company. Overall, the obtained results motivate us to continue further evolve our approach. In fact, the improvement will include (i) using a corpus of software-related texts for further pre-training, (ii) fine-tuning models on a multi-task datasets and (iii) annotating more data with requirements.

References

1. Ferrari, A., Spagnolo, G.O., Gnesi, S.: Pure: A dataset of public requirements documents, In: 2017 IEEE 25th International Requirements Engineering Conference (RE), pp. 502–505 (2017)
2. Devlin, J., Chang, M.-W., Lee, K., Toutanova, K.: BERT: Pre-training of deep bidirectional transformers for language understanding (2018). arXiv preprint arXiv:1810.04805
3. Abualhaija, S., Arora, C., Sabetzadeh, M., Briand, L.C., Vaz, E.: A machine learning-based approach for demarcating requirements in textual specifications. In: 2019 IEEE 27th International Requirements Engineering Conference (RE), pp. 51–62 (2019)
4. Bojanowski, P., Grave, E., Joulin, A., Mikolov, T.: Enriching word vectors with subword information. Trans. Assoc. Comput. Linguist. **5**, 135–146 (2017)
5. Peters, M.E., et al.: Deep contextualized word representations (2018). arXiv preprint arXiv:1802.05365
6. Sun, C., Qiu, X., Xu, Y., Huang, X.: How to fine-tune BERT for text classification? (2020)
7. Zhao, L., et al.: Natural language processing (NLP) for requirements engineering: A systematic mapping study (2020). arxiv.org/abs/2004.01099
8. Ghosh, S., Elenius, D., Li, W., Lincoln, P., Shankar, N., Steiner, W.: Arsenal: Automatic requirements specification extraction from natural language. In: Rayadurgam, S., Tkachuk, O. (eds.) NASA Formal Methods, pp. 41–46. Springer International Publishing, Cham (2016)

9. Körner, S.J., Landhäußer, M.: Semantic enriching of natural language texts with automatic thematic role annotation. In: Hopfe, C.J., Rezgui, Y., Métais, E., Preece, A., Li, H. (eds.) NLDB 2010. LNCS, vol. 6177, pp. 92–99. Springer, Heidelberg (2010). https://doi.org/10.1007/978-3-642-13881-2_9

10. Maiden, N., Lockerbie, J., Zachos, K., Bertolino, A., De Angelis, G., Lonetti, F.: A requirements-led approach for specifying qos-aware service choreographies: An experience report. In: Salinesi, C., van de Weerd, I. (eds.) Requirements Engineering: Foundation for Software Quality, pp. 239–253. Springer International Publishing, Cham (2014)

11. Zhong, H., Zhang, L., Xie, T., Mei, H.: Inferring resource specifications from natural language api documentation, In: IEEE/ACM International Conference on Automated Software Engineering, pp. 307–318 (2009)

12. Shakeri Hossein Abad, Z., Gervasi, V., Zowghi, D., Barker, K.: Elica: An automated tool for dynamic extraction of requirements relevant information. In: 2018 5th International Workshop on Artificial Intelligence for Requirements Engineering (AIRE), pp. 8–14 (2018)

13. Bakar, N.H., Kasirun, Z.M., Salleh, N., Jalab, H.A.: Extracting features from online software reviews to aid requirements reuse. In: Applied Soft Computing, vol. 49, pp. 1297–1315 (2016). www.sciencedirect.com/science/article/pii/S1568494616303830

14. Nguyen, T.H., Grundy, J., Almorsy, M.: Rule-based extraction of goal-use case models from text. In: Proceedings of the 2015 10th Joint Meeting on Foundations of Software Engineering, ser. ESEC/FSE 2015. New York, NY, USA: Association for Computing Machinery, pp. 591–601 (2015). https://doi.org/10.1145/2786805.2786876

15. Zeni, N., Kiyavitskaya, N., Mich, L., Cordy, J.R., Mylopoulos, J.: Gaiust: supporting the extraction of rights and obligations for regulatory compliance. Requir. Eng. **20**(1), 1–22 (2015)

16. Casagrande, E., Woldeamlak, S., Woon, W.L., Zeineldin, H.H., Svetinovic, D.: NLP-KAOS for systems goal elicitation: smart metering system case study. IEEE Trans. Softw. Eng. **40**(10), 941–956 (2014)

17. Johann, T., Stanik, C., Alizadeh B, A.M., Maalej, W.: Safe: a simple approach for feature extraction from app descriptions and app reviews. In: 2017 IEEE 25th International Requirements Engineering Conference (RE), pp. 21–30 (2017)

18. Zhang, J., El-Gohary, N.M.: Integrating semantic NLP and logic reasoning into a unified system for fully-automated code checking. In: Automation in Construction, vol. 73, pp. 45–57 (2017). www.sciencedirect.com/science/article/pii/S0926580516301819

19. Xiao, X., Paradkar, A., Thummalapenta, S., Xie, T.: Automated extraction of security policies from natural-language software documents. In: Proceedings of the ACM SIGSOFT 20th International Symposium on the Foundations of Software Engineering, ser. FSE '12. New York, NY, USA: Association for Computing Machinery (2012). https://doi.org/10.1145/2393596.2393608

20. Ochodek, M., Nawrocki, J.: Automatic transactions identification in use cases. In: Meyer, B., Nawrocki, J..R.., Walter, B. (eds.) CEE-SET 2007. LNCS, vol. 5082, pp. 55–68. Springer, Heidelberg (2008). https://doi.org/10.1007/978-3-540-85279-7_5

21. Rolland, C., Achour, C.B.: Guiding the construction of textual use case specifications," Data & Knowledge Engineering, vol. 25, no. 1, pp. 125–160 (1998). www.sciencedirect.com/science/article/pii/S0169023X97862234

22. Tiwari, S., Ameta, D., Banerjee, A.: An approach to identify use case scenarios from textual requirements specification. In: Proceedings of the 12th Innovations on Software Engineering Conference (Formerly Known as India Software Engineering Conference), Ser. ISEC'19. New York, NY, USA. Association for Computing Machinery (2019). https://doi.org/10.1145/3299771.3299774
23. Hey, T., Keim, J., Koziolek, A., Tichy, W.F.: Norbert: transfer learning for requirements classification. In: 2020 IEEE 28th International Requirements Engineering Conference (RE), pp. 169–179 (2020)

Continuous Prompt Tuning for Russian: How to Learn Prompts Efficiently with RuGPT3?

Nikita Konodyuk[1,2] and Maria Tikhonova[1,2(✉)]

[1] SberDevices, Sberbank, Moscow, Russia
`tikhonova.m.iva@sberbank.ru`
[2] National Research University Higher School of Economics, Moscow, Russia
`nekonodyuk@edu.hse.ru`

Abstract. Adaptation to downstream tasks is a crucial part of the pre-trained language model (PLM) life cycle. Fine-tuning, traditionally used for this purpose, is an expensive procedure in terms of computation and memory. Dramatic growth of PLM capacities has led to the emergence of zero- and few-shot methods, which use natural language to describe tasks. Although these methods do not modify the parameters of the model, they rely on manual prompt design, which may be suboptimal. To address this issue, a range of techniques for automatic prompt search have been proposed recently.

In this paper, we present a framework for continuous prompt tuning (CPT) in Russian. We evaluated our framework by adapting RuGPT3 to tasks in the Russian benchmark SuperGLUE. We obtained metrics better or comparable to fine-tuning, while training only an auxiliary model that provides prompt embeddings, so the total number of trained parameters accounts for less than 0.4% of that of RuGPT3. In addition, we conducted experiments comparing different configurations of the framework and explored the lower bound to which we can reduce the number of parameters. Our source code is publicly available at
https://github.com/sberbank-ai/ru-prompts.

Keywords: Natural language processing · Language models · Model training · Transformer models · Language model adaptation

1 Introduction

Language models, in particular, Generative Pre-trained Transformers, have shown prominent abilities for many Natural Language Processing (NLP) tasks. The pre-training fine-tuning paradigm for solving downstream tasks [1], which has been the dominant approach, especially for transformer models, will be limited as long as it requires large labelled training corpora. Moreover, fine-tuning large language models can be computationally expensive and time-consuming.

In [2] the authors present GPT3, an autoregressive language model, which can be applied without any gradient updates or fine-tuning, with tasks and

E. Burnaev et al. (Eds.): AIST 2021, CCIS 1573, pp. 30–40, 2022.
https://doi.org/10.1007/978-3-031-15168-2_3

few-shot demonstrations specified purely via text interaction with the model. The methodology in [3] introduces the concept of **few-shot**: the model receives several training examples and a test prompt in a text format as an input and makes predictions based on them. The setting when the model receives no training examples and makes a prediction based only on the text prompt constructed from a test sample is called zero-shot. To illustrate the idea of a text prompt we present an example for the DaNetQA task[1] (a question answering task for questions with binary answers *yes* or *no* from the Russian benchmark SuperGLUE) in Fig. 1. After the format of the text prompt is defined, we sequentially unite it with each of the answer options (that is, each of the possible labels) and measure the perplexity of the resulting text fragment. The answer options are then sorted by the perplexity scores, and the one with the lowest score is considered to be a prediction. Thus, such an approach does not require any additional training and the answer can be obtained with the use of the original pre-trained model.

```
test_sample = {
                "question": "А была ли блокада Ленинграда?",
                "passage": "7 февраля — в блокадный Ленинград
                прибыл первый поезд с «Большой земли» [...]
                27 января: Снята блокада Ленинграда.
                "idx": 14
            }

prompt      =     passage + " \n" + question
```

Fig. 1. A text prompt constructed for a DaNetQA test sample.

These few-shot and zero-shot methods have shown promising results on a wide range on Natural Language Processing (NLP) tasks, especially with large generative models. However, such an approach obviously has disadvantages, one of the most important of which is that even a slight change in the prompt format may significantly influence the result. Since manual selection of prompt templates is not optimal, the idea naturally arises: to automatically search prompts. Since recently methods which tune prompt-embedding in discrete and continuous spaces have been actively developing.

Several attempts made in this direction focused on discrete prompt search [4–6] and demonstrated the effectiveness of an automated approach. However, as long as neural language models are inherently continuous, discrete search for the prompts is likely to be sub-optimal. Thus, the next step is to search pseudo-prompts in continuous embedding space. This idea has been explored in several recent works [7–10] and has proved to be quite fruitful.

---

[1] https://russiansuperglue.com/tasks/task_info/DaNetQA.

In this work we follow the idea of the P-tuning method introduced in [11], where the authors use a bidirectional LSTM to learn continuous prompt embeddings. Namely, we present a framework for *Continuous Prompt Tuning* (or simply CPT) for the Russian language. In addition, we carry out a series of experiments on the Russian SuperGLUE benchmark [12][2] comparing CPT with zero-shot and standard fine-tuning. We explore the influence of the number of trainable parameters on the result. The code is publicly available in our GitHub repository.[3]

Thus, the contribution of this work is three-fold: (i) we release the framework for CPT for Russian, which can be easily adapted to various models; (ii) we evaluate CPT for the RuGPT3 model on Russian SuperGLUE and show that it can be regarded as a strong competitor to fine-tuning and zero-shot; (iii) we show that the number of trainable LSTM parameters can be reduced without significant losses in total quality.

This paper is structured as follows: Sect. 2 describes the method implemented in the CPT framework; Sect. 3 presents the evaluation setup and the analysis of the conducted experiments; Sect. 4 is devoted to the analysis of the model behavior and discusses the results; and, finally, Sect. 5 concludes the paper.

2   Method

In this section we introduce CPT, deriving it from few-shot and zero-shot settings as they are introduced in [2]. We consider a classification task and follow the *generative classification* paradigm, where prediction for a prompt is inferred from the first token generated by a model after processing the prompt.

A natural language prompt is the core element of all prompt-based methods. It combines a description of a downstream task with optional examples that should also be given in the format, which should be clearly understood by the model. Manual prompt search is always a matter of trial and error and therefore is substantially hard to formalize. Nevertheless, in the vast majority of cases, the prompt consists of the same set of semantic blocks.

To illustrate this idea, let us consider the following example of a few-shot prompt for a machine translation task:

> Translate English to French
> sea otter => loutre de mer
> plush giraffe => girafe peluche
> cheese => (MASK)

In this case the prompt takes the following formal format:

---

[2] https://russiansuperglue.com/.
[3] https://github.com/sberbank-ai/ru-prompts.

> Translate English to French
> {word_in_english} => {word_in_french}
> {word_in_english} => {word_in_french}
> {word_in_english} => (MASK)

In the example above we have pairs of instance queries (or objects) and targets , where the answer for the last instance is masked, as well as special service elements which we further refer to as task instructions (or simply TI), which help the model understand what is required in the task. In fact, in other applications we also encounter such elements as task context and instance context . They are necessary in such tasks as summarization, i.e. in those where the query depends on additional context.

Thus, the generic few-shot prompt format can be formalized as follows:

> ⟨TI⟩ ⟨task context⟩ ⟨TI⟩
> ⟨TI⟩ ⟨instance context⟩ ⟨TI⟩ ⟨instance query⟩ ⟨TI⟩ ⟨instance target⟩
> ⟨TI⟩ ⟨instance context⟩ ⟨TI⟩ ⟨instance query⟩ ⟨TI⟩ ⟨instance target⟩
> ⟨TI⟩ ⟨instance context⟩ ⟨TI⟩ ⟨instance query⟩ ⟨TI⟩ (MASK)

Since the zero-shot approach differs from the few-shot only in terms of the number of provided examples, the zero-shot prompt format in essence is just a truncation of the few-shot prompt format:

> ⟨TI⟩ ⟨instance context⟩ ⟨TI⟩ ⟨instance query⟩ ⟨TI⟩ (MASK)

For example:

> Известно, что Москва была основана в 1147 году на Москве-реке.
> Вопрос: Была ли Москва основана в 12 веке? Ответ: (MASK)

Note that everything except the task instruction is usually derived from the dataset fields and thus is not subject to change. Task instructions, on the contrary, are defined by the task itself and even minor changes lead to significant performance deviations. Moreover, for different models different TI may be optimal. Although hard to discover manually, they can be trained by gradient descent, as proposed in [8,9,11] in different variations.

We train task instructions by gradient descent, so the prompt takes the following form:

> ⟨learned instructions⟩ Москва была основана в 1147 году на Москве-реке.
> ⟨learned instructions⟩ Была ли Москва основана в 12 веке?
> ⟨learned instructions⟩ (MASK)

We follow the original methodology proposed in [11] and produce trainable embeddings with an auxiliary BiLSTM-based model, which we also refer to as *prompt provider*. Its architecture is as follows: a sequence of trainable vectors is passed through BiLSTM and then through two-layer MLP with ReLU activation. The dimension of the output sequence of vectors is equal to the embedding dimension of backbone model, and their number is equal to the total number of ⟨SP⟩ tokens in *prompt format*. These embeddings are inserted to the corresponding positions in the input of backbone and trained via backpropagation.

3   Experiments

In this section, we describe experiments conducted on the Russian SuperGLUE benchmark. First, we give a brief description of the tasks on which we evaluated CPT, then we describe the model used in the experiments, after that we specify details about the baseline methods and, finally, present the results and their analysis. We conclude the section with an additional series of experiments with a different number of trainable parameters.

3.1   Data

All the experiments were conducted using the Russian general language understanding evaluation benchmark – RussianGLUE. It was collected and organized analogically to the SuperGLUE methodology [13]. Russian SuperGLUE comprises 9 tasks divided into 5 groups:

- **Textual Entailment & Natural Language Inference (NLI)**: TERRa, RCB, LiDiRus;
- **Common Sense**: RUSSE, PARus;
- **World Knowledge**: DaNetQA;
- **Machine Reading**: MuSeRC, RuCoS;
- **Reasoning**: RWSD.

Below a brief description of each task is given, and aggregated information is presented in Table 1.

TERRA Textual Entailment Recognition for Russian is aimed at capturing textual entailment in a binary classification form. Given two text fragments (premise and hypothesis), the task is to determine whether the meaning of the hypothesis is entailed from the premise. The dataset was sampled from the Taiga corpus [14].

RCB The Russian Commitment Bank is a 3-way classification task aimed at recognizing textual entailment (NLI). Analogically to TERRA, each example in RCB consists of premise and hypothesis. However, in this task a premise can be a short paragraph, not necessarily one phrase.

LiDiRus (also referred to as a diagnostic dataset) is an expert-constructed evaluation dataset for recognizing textual entailment tasks on paired sentences. It is a direct translation from the English SuperGLUE diagnostic dataset, originally introduced in [15]. It consists of 1104 sentence pairs which are used as a test set for testing models' capacity to solve NLI task. In addition, the diagnostic dataset has a rich annotation of various linguistic phenomena, partly inserted artificially to explore the possible biases and errors on a task that can be considered truly universal for all languages. The annotation includes 33 features which can be devided into 4 categories: *Predicate-Argument Structure, Logic, Lexical-Semantics, and Knowledge*. Such annotation makes it possible to analyze the model behaviour with respect to linguistic features.

RUSSE is a binary classification task that involves word sense disambiguation. Given a pair of sentences containing the same ambiguous word, the goal of the model is to recognize if the word is used in the same meaning. The dataset was constructed from RUSSE[4].

PARus is a binary classification task aimed at identifying the most plausible alternative out of two for a given premise. The correct alternatives is the dataset are randomized so that the expected performance of random guessing yields 50% accuracy score.

DaNetQA is a Russian question-answering dataset for questions with binary answers (*yes* or *no*) which follows the BoolQ design. Each example consists of a triplet of question, passage, and answer.

MuSeRC is a machine reading comprehension (MRC) task. Each sample consists of a text paragraph, multi-hop questions based on the paragraph, and possible answers for each question. The goal of the task is to choose all correct answers for each question.

Table 1. Russian SuperGLUE task description. Train/Val/Test stand for example amount (sentence pairs or texts); MCC stands for Matthews Correlation Coefficient; EM - Exact Match.

Task	Task type	Task metric	Train	Val	Test
TERRa	NLI	Accuracy	2616	307	3198
RCB	NLI	Avg. F1/Accuracy	438	220	438
LiDiRus	NLI & Diagnostics	MCC	0	0	1104
RUSSE	Common sense	Accuracy	19845	8508	18892
PARus	Common sense	Accuracy	400	100	500
DaNetQA	World knowledge	Accuracy	1749	821	805
MuSeRC	Machine reading	F1/EM	500	100	322
RuCoS	Machine reading	F1/EM	72193	7577	7257
RWSD	Reasoning	Accuracy	606	204	154

RuCoS is an MRC task that involves commonsense reasoning and world knowledge. The dataset is a counterpart of ReCoRD[5] for English.

RWSD The Russian Winograd Schema task is devoted to coreference resolution in a binary classification form. The corpus was created as a manually validated translation of the Winograd Schema Challenge[6].

---

[4] https://russe.nlpub.org/downloads/.

[5] https://sheng-z.github.io/ReCoRD-explorer/.

[6] https://cs.nyu.edu/faculty/davise/papers/WinogradSchemas/WS.html.

3.2   Model

We investigate the effectiveness of CPT and conduct all the experiments using a RuGPT3[7] model, a Russian adaptation of the autoregressive language model GPT3 [2], which the authors claimed as having strong in-context learning abilities and which has shown impressive results in zero- and few-shot settings in many NLP tasks. Namely, we run the experiments on *RuGPT3-Large* which is publicly available in the Hugging Face Python library[8].

For a prompt provider we used an LSTM with a hidden dimension of 256 and input dimension of 16. Thus, given that the number of parameters of RuGPT3-Large equals 760M and the number of trainable parameters of the prompt provider was 3M, the fraction of total trainable parameters accounts only for 0.4% of model size. Additionally, in Sect. 3.5 we experiment with different numbers of parameters in the prompt provider.

We cast each of the tasks described in Sect. 3.1 to a binary or ternary classification problem. Training CPT on a task involves formatting its samples in line with the corresponding prompt format from Table 2, where ⟨target⟩ takes the values of "да" and "нет" (which are also called *label verbalizers*), and additionally "возможно" in the case of ternary classification. The symbol ⟨SP⟩ represents a trainable token (soft prompt). Following the original methodology from [11] we used sequences of 3 soft prompt tokens. However, it should be noted that the number of soft prompt tokens can be varied and the optimal choice of the number of ⟨SP⟩ in each sequence for each task is another area for the research. The embeddings of trainable tokens are provided by the prompt provider. We train the prompt provider to output such embeddings, which maximize the probability of the right label verbalizer for each sample. As a criterion we use cross entropy among the logits corresponding to the label verbalizers.

Table 2. Prompt formats used for training on Russian SuperGLUE tasks. The symbol ⟨SP⟩ represents a trainable token (soft prompt).

Task	Prompt Format
DaNetQA	⟨SP⟩⟨SP⟩⟨SP⟩ ⟨passage⟩ ⟨SP⟩⟨SP⟩⟨SP⟩ ⟨question⟩ ⟨SP⟩⟨SP⟩⟨SP⟩ ⟨target⟩
TERRa	⟨SP⟩⟨SP⟩⟨SP⟩ ⟨premise⟩ ⟨SP⟩⟨SP⟩⟨SP⟩ ⟨hypothesis⟩ ⟨SP⟩⟨SP⟩⟨SP⟩ ⟨target⟩
LiDiRus	⟨SP⟩⟨SP⟩⟨SP⟩ ⟨sentence1⟩ ⟨SP⟩⟨SP⟩⟨SP⟩ ⟨sentence2⟩ ⟨SP⟩⟨SP⟩⟨SP⟩ ⟨target⟩
MuSeRC	⟨SP⟩⟨SP⟩⟨SP⟩ ⟨paragraph⟩ ⟨SP⟩⟨SP⟩⟨SP⟩ ⟨question⟩ ⟨SP⟩⟨SP⟩⟨SP⟩ ⟨answer⟩ ⟨SP⟩⟨SP⟩⟨SP⟩ ⟨target⟩
PARus	⟨SP⟩⟨SP⟩⟨SP⟩ ⟨premise⟩ ⟨SP⟩⟨SP⟩⟨SP⟩ ⟨choice2⟩ ⟨SP⟩⟨SP⟩⟨SP⟩ ⟨choice1⟩ ⟨SP⟩⟨SP⟩⟨SP⟩ ⟨target⟩
RCB	⟨SP⟩⟨SP⟩⟨SP⟩ ⟨premise⟩ ⟨SP⟩⟨SP⟩⟨SP⟩ ⟨hypothesis⟩ ⟨SP⟩⟨SP⟩⟨SP⟩ ⟨target⟩
RUSSE	⟨SP⟩⟨SP⟩⟨SP⟩ ⟨word⟩ ⟨SP⟩⟨SP⟩⟨SP⟩ ⟨sentence1⟩ ⟨SP⟩⟨SP⟩⟨SP⟩ ⟨sentence2⟩ ⟨SP⟩⟨SP⟩⟨SP⟩ ⟨target⟩
RWSD	⟨SP⟩⟨SP⟩⟨SP⟩ ⟨text⟩ ⟨SP⟩⟨SP⟩⟨SP⟩ ⟨span1_text⟩ ⟨SP⟩⟨SP⟩⟨SP⟩ ⟨span2_text⟩ ⟨SP⟩⟨SP⟩⟨SP⟩ ⟨target⟩
RuCoS	⟨SP⟩⟨SP⟩⟨SP⟩ ⟨passage⟩ ⟨SP⟩⟨SP⟩⟨SP⟩ ⟨statement⟩ ⟨SP⟩⟨SP⟩⟨SP⟩ ⟨target⟩

---

[7] https://github.com/sberbank-ai/ru-gpts.

[8] https://huggingface.co/sberbank-ai/rugpt3large_based_on_gpt2.

3.3 Baselines

In order to evaluate the proposed framework we conducted a series of experiments with our RuGPT3-Large model, comparing CPT with standard fine-tuning and zero-shot approaches.

We fine-tuned RuGPT3-Large for every task using *jiant-russian* (version 2.0) library[9] (a library released by the creators of the benchmark, which is aimed at fine-tuning various models on Russian SuperGLUE) with standard parameter configuration.

For the zero-shot method we used its modification based on the model perplexity. Namely, for each test sample we calculate the perplexity of the corresponding prompts united with one of the possible targets using formula 1. We then choose the best target, as the one with the lowest perplexity score.

$$PPL(t) = \exp\left(-\frac{1}{|t|} \sum_{i=0}^{|t|} \log_{p_\theta}(x_i|x_{<i})\right) \tag{1}$$

where t is an input text (in our case, a text prompt concatenated with one of possible targets), $|t|$ is the length of the text in tokens, and $log_{p_\theta}(x_i|x_{<i})$ is the log-likelihood of the i-th token in t conditioned on the preceding ones.

3.4 Results

The results of the experiments are presented in Table 3. It gives an exact representation of CPT performance compared with zero-shot and fine-tuning approaches. It can be seen that CPT outperforms each of zero-shot and fine-tuning on most of the tasks. Namely, it outperforms zero-shot on LiDiRus, MuSeRC, TERRa, RUSSE, RWSD, and DaNetQA; and it shows better results than fine-tuning on MuSeRC, TERRa, RWSD, DaNetQA and RuCoS. Thus, CPT allows to achieve reasonable model performance without either human assistance in prompt search or computational resources sufficient for fine-tuning.

The poor performance of CPT on RCB can be explained by the small size of the training corpora (only 438 training samples), which is insufficient for learning good prompts. As for RuCoS, the most plausible explanation is that the chosen generative approach is not optimal for such a complicated type of task. In the future we plan to use CPT with contrastive classification for this task (see Sect. 4 for a more in-depth description of the approach), which will hopefully yield better results.

3.5 Experiments with Different Number of Trainable Parameters

In addition, we explored how the number of trainable parameters in an LSTM influence the model quality. Our goal was to minimize the number of trainable

---

[9] https://github.com/RussianNLP/RussianSuperGLUE/tree/master/jiant-russian-v2.

Table 3. Results of RuGPT3-Large evaluation on Russian SuperGLUE in different settings. We score the tasks in line with the metrics specified in Table 1. The scores for all tasks are then averaged to get the total score. For the tasks with multiple metrics, the metrics are averaged.

Approach	Total score	LiDiRus	RCB	PARus	MuSeRC	TERRa	RUSSE	RWSD	DaNetQA	RuCoS
Zero-shot	51.4	12.8	30.4/42.2	63.0	72.7/52.2	52.5	57.1	62.3	57.0	64.0/63.5
Fine-tuning	50.5	23.1	41.7/48.4	58.4	72.9/33.3	65.4	64.7	63.6	60.4	21.0/20.2
CPT	48.2	14.0	17.6/35.8	47.2	74.2/38.3	67.9	62.8	66.9	60.7	32.0/31.4

parameters and, therefore, to optimize and speed up CPT training. For this purpose we conducted a series of experiments on 4 Russian SuperGLUE tasks (DaNetQA, PARus, RCB, and TERRa). These tasks were chosen as long as they are considered to be most popular among all the benchmark tasks and due to the size of their training corpora.

In the experiments we trained CPT clones, each with a different dimension of LSTM hidden states varying from 1 to 1536. Results are presented in Table 4 and a general picture is given in Fig. 2. It can be seen that the number of the hidden dimensions and, therefore, the number of the trainable parameters can be significantly reduced without noticeable decrease in total quality. Moreover, an excessive number of trainable parameters may seemingly lead to overfitting and a non-optimal score. For instance, while for TERRa and DaNetQA the accuracy keeps increasing, on PARus it reaches the optimal value on `hidden_dim=16` and then decreases. We connect this behaviour with the number of training samples, which is significantly greater for DaNetQA and TERRa, than for PARus and RCB.

If we calculate the average of the scores for the 4 tasks considered, we see that the result for `hidden_dim=64` is only 7,5% worse than for the maximal size of `hidden_dim=1536` while it requires 368 times fewer trainable parameters. Thus, it can be concluded that the number of trainable parameters can be significantly reduced without much loss in quality.

Table 4. CPT results with different LSTM hidden dimensions. "Hidden dim" stands for the number of LSTM hidden dimensions and "Params" for the number of trainable parameters. The average score is calculated as the mean score of 4 tasks. For RCB the two metrics are first averaged.

hidden_dim	Params	DaNetQA	PARus	RCB	TERRa	Average
1	3.2K	50.3	48.8	17.6/35.6	50.3	44.0
4	8.5K	53.4	47.0	20.7/32.9	50.3	44.4
16	37.5K	52.0	53.0	17.6/35.8	53.8	46.4
64	274K	54.7	51.2	17.5/35.6	54.8	46.8
256	3.2M	56.9	47.4	17.6/35.8	53.2	46.1
1024	45.7M	58.0	48.4	17.6/35.8	60.4	48.4
1536	101M	60.7	47.2	17.6/35.8	67.9	50.6

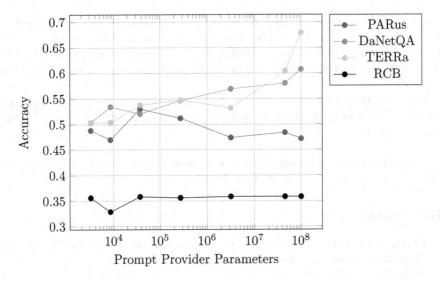

Fig. 2. Accuracy score for CPT with different LSTM hidden dimensions.

4   Discussion

Despite the fact that the overall RuGPT3 performance with CPT is comparable with fine-tuning and zero-shot approaches, it shows quite a poor score on several tasks (namely, RCB, PARus, and RuCoS). This may be explained in several ways. For example, we suppose that the low score on RCB and PARus can be accounted for by the small size of the training corpora (438 and 400 training samples respectively). Such a modest dataset size is probably not enough for learning good continuous prompts.

As for RuCoS, its low score can be explained by the fact that a generative approach is not optimal for such a complicated task. Thus, in order to overcome this limitation we are planning to use CPT for contrastive classification. Compared with generative classification, this approach will utilize the relationships of multiple versions of each text, for example multiple answers or multiple prompts, thus being able to handle multiple-choice tasks in a more natural manner. Another use case of contrastive classification will probably be multiclass classification where it is hard to choose suitable label verbalizers.

Another area for future research could become imposing additional restrictions on prompt provider to increase interpretability of the output prompt. Although they are practically efficient, they currently remain non-interpretable.

5   Conclusion

In this paper we propose a framework for continuous prompt tuning for the Russian language. We use an RuGPT3 model and evaluate it on the Russian

SuperGLUE benchmark. In the experiments our method shows results competitive with zero-shot and fine-tuning and even outperforms them on most of the tasks. In addition, we explore the influence of the number of trainable LSTM parameters and find out that it can be significantly reduced without any losses in quality.

In the future we are planning to implement a contrastive classification approach for CPT and apply the framework to other models, such as the RuT5 and RuBERT models.

Acknowledgements. We would like to thank Sarah Caitlin Bennett for her help with editing the paper and advice on the text structure.

References

1. Devlin, J., Chang, M.-W., Lee, K., Toutanova, K.: Bert: pre-training of deep bidirectional transformers for language understanding. arXiv preprint arXiv:1810.04805 (2018)
2. Brown, T.B., et al.: Language models are few-shot learners. arXiv preprint arXiv:2005.14165 (2020)
3. Radford, A., Jeffrey, W., Child, R., Luan, D., Amodei, D., Sutskever, I.: Language models are unsupervised multitask learners. OpenAI Blog **1**(8), 9 (2019)
4. Shin, T., Razeghi, Y., Logan IV, R.L., Wallace, E., Singh, S.: Autoprompt: eliciting knowledge from language models with automatically generated prompts. arXiv preprint arXiv:2010.15980 (2020)
5. Reynolds, L., McDonell, K.: Prompt programming for large language models: beyond the few-shot paradigm. In: Extended Abstracts of the 2021 CHI Conference on Human Factors in Computing Systems, pp. 1–7 (2021)
6. Gao, T., Fisch, A., Chen, D.: Making pre-trained language models better few-shot learners. arXiv preprint arXiv:2012.15723 (2020)
7. Li, X.L., Liang, P.: Prefix-tuning: optimizing continuous prompts for generation. arXiv preprint arXiv:2101.00190 (2021)
8. Lester, B., Al-Rfou, R., Constant, N.: The power of scale for parameter-efficient prompt tuning. arXiv preprint arXiv:2104.08691 (2021)
9. Hambardzumyan, K., Khachatrian, H., May, J.: Warp: word-level adversarial reprogramming. arXiv preprint arXiv:2101.00121 (2021)
10. Liu, X., Ji, K., Fu, Y., Du, Z., Yang, Z., Tang, J.: P-tuning v2: prompt tuning can be comparable to fine-tuning universally across scales and tasks. arXiv preprint arXiv:2110.07602 (2021)
11. Liu, X., et al.: GPT understands, too. arXiv preprint arXiv:2103.10385 (2021)
12. Shavrina, T., et al.: Russiansuperglue: a Russian language understanding evaluation benchmark. arXiv preprint arXiv:2010.15925 (2020)
13. Wang, A., et al.: Superglue: a stickier benchmark for general-purpose language understanding systems. arXiv preprint arXiv:1905.00537 (2019)
14. Shavrina, T., Shapovalova, O.: To the methodology of corpus construction for machine learning: "taiga" syntax tree corpus and parser. In: Proceedings of "CORPORA-2017" International Conference, pp. 78–84 (2017)
15. Wang, A., Singh, A., Michael, J., Hill, F., Levy, O., Bowman, S.R.: Glue: a multitask benchmark and analysis platform for natural language understanding. arXiv preprint arXiv:1804.07461 (2018)

An Intelligent Web-Service for Automatic Concept Map Generation

Aliya Nugumanova[1]($\boxtimes$) (iD), Yerzhan Baiburin[1] (iD), and Kurmash Apayev[2] (iD)

[1] Sarsen Amanzholov East Kazakhstan University, Shakarim 148,
070000 Ust-Kamenogorsk, Kazakhstan
yalisha@yandex.kz

[2] D. Serikbayev East Kazakhstan Technical University, Serikbayev 19,
070000 Ust-Kamenogorsk, Kazakhstan

Abstract. Automatic generation of concept maps from texts is yet another research challenge at the intersection of Natural Language Processing (NLP) and Knowledge Engineering, that has not been fully resolved. Concept maps are knowledge visualization tools representing texts at the conceptual level. They are intended to show systemic relations between key concepts of a given text and contribute to its deeper understanding and appreciation. In this work, we present a web service for automatic generation of concept maps. The service provides both a simple web interface and API access. It takes an input text in Kazakh, Russian or English languages and generates a concept map, using one of two existing algorithms at the user's choice. Our service in a sense eliminates the need for human effort in creating concept maps, and can be useful in teaching, developing research, and exploring texts. In addition to the proposed web service, the contribution of our paper consists in the adaptation of the used algorithms originally developed for texts in English, to Kazakh, which is recognized as low resource language in NLP.

Keywords: Concept map mining · Automatic generation of concept maps · Knowledge graph · Relation extraction · Web-service

1 Introduction

Concept maps play important role in education and conceptualization [1]. They serve as cognitive visualization tools along with mind maps, causal diagrams, Petri nets, knowledge graphs etc. The latter (knowledge graphs) are structurally indistinguishable from concept maps and represented in the same manner – through nodes (entities) and edges (semantic relations). Compared to concept maps, knowledge graphs can express more extensive knowledge and provide more comprehensive functions; for this reason, they are more automated and contribute to many applications such as search engines, recommender systems, AI assistants etc. [2, 3]. Well-known examples of knowledge graphs are DBpedia, Freebase and YAGO [4]. Nowadays, the "great migration" from old-school Freebase to Google Knowledge Graph and WikiData is complete [5, 6].

© The Author(s), under exclusive license to Springer Nature Switzerland AG 2022
E. Burnaev et al. (Eds.): AIST 2021, CCIS 1573, pp. 41–44, 2022.
https://doi.org/10.1007/978-3-031-15168-2_4

In this work, we explore the process of construction of concept maps, but consider methods developed for knowledge graphs. We have two reasons for this. Firstly, there a lot in common between these two processes. Both processes involve the stage of knowledge extraction, while the stages of knowledge storing and sharing, are specific only to construction of knowledge graphs. At second, knowledge graphs are currently one of the most popular formats for representing knowledge in the subject areas of Industry 4.0 [7], due to which methods for their construction are more developed. There are two main paradigms for constructing knowledge graphs: manual construction by experts, and automatic extraction from trusted sources, including the textual content of web pages [8]. Recently with the surge of deep learning models [9], many approaches have been developed for automatic generation of knowledge graphs from texts. For example, in [10], the BERT Transformer model is fine-tuned on COVID-19 texts and then applied to generate the COVID-19 knowledge graph. Transformer-based models such as BERT, RoBERTa, and XLNet are widely used for relation extraction and knowledge graph generation in many domain areas, e.g. medicine [11], agriculture [12], chemistry [13].

2   Methodology

In this Section, we propose a web service for automatically generating concept maps from textual documents. The service allows to easily access and consume two knowledge graph generation algorithms through Web API. Both algorithms utilize recent advances in pre-trained language models, and provide Kazakh, Russian and English languages. The first algorithm extends ReVerb method [14] by introducing pretrained language models from the SpaCy library. The second algorithm exploits ideas from [15] and uses the BERT language model.

2.1   ReVerb Algorithm for Automatic Concept Map Generation

ReVerb first identifies relation phrases that satisfy the syntactic and lexical constraints, and then finds a pair of entities (noun-phrase arguments) for each identified relation phrase [14]. This method only extracts verbal relation sequences between two entities tokens, e.g., relation sequences such as "We trust in God", while relation sequencies such as in "In God we trust" are ignored. The summary of relation extraction procedure is as follows. For each verb v in a sentence s, find the longest sequence of words r_v such that: (1) r_v starts at the verb v; (2) r_v satisfies the syntactic constraint (matches POS-tag patterns); (3) r_v satisfies the lexical constraint (database is used). If any pair of verbal sequences r_{v1} and r_{v2} are adjacent or overlap in the sentence s, they are merged into a single sequence.

The summary of entity extraction procedure is as follows. For each identified relation phrase r: (1) find the nearest noun phrase x to the left of r in a sentence s such that x is not a relative pronoun, WHO-adverb, or existential "there"; (2) find the nearest noun phrase y to the right of r in a sentence s. If such an (x, y) pair could be found, return (x, r, y) as an extraction.

We realize ReVerb method with the help of SpaCy, a Python NLP library. SpaCy provides pre-trained language models (built-in and third-party) for various languages. Pretraining means that we do not need to construct a relation database, we can use patterns and form queries to the pretrained language model.

2.2 "Matching the Blanks" (MTB) Algorithm for Automatic Concept Map Generation

This method is focused on mapping relation statements to relation representations using BERT model [15]. Let $x = [x_0 \ldots x_n]$ be a sequence of tokens, where $x_0 = [CLS]$ and $x_n = [SEP]$ are tags for start and end of a sequence. Let $s_1 = (i, j)$ and $s_2 = (k, l)$ – are pairs of integers, where $i < j < k \leq l \leq n$. A relation statement is a triple $r = (x, s_1, s_2)$, where the indices s_1 and s_2 separate entity mentions in x, i.e., the sequence $\left[x_i \ldots x_{j-1}\right]$ mentions the head entity, and the sequence $\left[x_k \ldots x_{l-1}\right]$ mentions the tail entity. The goal is to learn a map function $f_\theta(r)$ that reflects the relation statement r to a fixed-length vector $h_r \in R^d$ that represents the relation expressed in x between the entities marked by s_1 and s_2.

There are several ways to represent entities in the input and a relation in the output of the model. We use the Entity Markers input representation and Entity Start output representation. We create pre-training data automatically using a predefined set of target entities for "Rivers" domain and then extract sentences from Wikipedia, which contain target entities. Then we annotate a small training dataset manually, and fine-tune BERT with this dataset for 10 epochs.

2.3 Web Service

The constructed web service for automatic concept map generation is available following URL: https://sorge.ektu.kz/api/?url=/api/maps.json#/. The service has three parameters: (1) a language (Kazakh, Russian or English), (2) a method for relation extraction (ReVerb referred to as "Rule-based", MTB referred to as "Deep Learning") and (3) an input text. The web service provides a simple user interface (UI) for interactive generation of concept map (see Fig. 1).

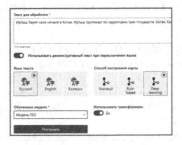

Fig. 1. Screenshot of the client interface connected to the web service through a browser

3 Conclusion and Future Work

In this paper, we presented the web service for automatic generation of concept maps, realizing two principal different algorithms: rule-based and deep-learning. Furthermore, a lightweight API was presented which allows to consume these algorithms online. We hope that our work eases the usage of concept maps in educational applications. In our future work, we plan to evaluate both used algorithms.

References

1. Novak, J.D., Cañas, A.: Theoretical origins of concept maps, how to construct them, and uses in education. Reflecting Educ. **3**(1), 29–42 (2007)
2. Liao, Y., et al.: Knowledge graph and its applications in MOOC and SPOC. In: 2019 2nd International Conference on Contemporary Education and Economic Development (CEED), pp. 301–305 (2019)
3. Goyal, N., et al.: CON2KG - a large-scale domain-specific knowledge graph. In: Proceedings of the 30th ACM Conference on Hypertext and Social Media, pp. 287–288 (2019)
4. Kertkeidkachorn, N., Ichise, R.: An automatic knowledge graph creation framework from natural language text. IEICE Trans. Inf. Syst. **101**(1), 90–98 (2018)
5. Pellissier Tanon, T., Vrandečić, D., Schaffert, S., Steiner, T., Pintscher, L.: From freebase to Wikidata: the great migration. In: Proceedings of the 25th International Conference on WWW, pp. 1419–1428 (2016)
6. Tiddi, I., Lécué, F., Hitzler, P. (eds.): Knowledge Graphs for Explainable Artificial Intelligence: Foundations, Applications and Challenges, vol. 47. IOS Press, Amsterdam (2020)
7. Bader, S.R., Grangel-Gonzalez, I., Nanjappa, P., Vidal, M.-E., Maleshkova, M.: A knowledge graph for Industry 4.0. In: Harth, A., et al. (eds.) ESWC 2020. LNCS, vol. 12123, pp. 465–480. Springer, Cham (2020). https://doi.org/10.1007/978-3-030-49461-2_27
8. Razniewski, S., Suchanek, F., Nutt, W.: But what do we know? In: Proceedings of the 5th Workshop on Automated Knowledge Base Construction, pp. 40–44 (2016)
9. Vaswani, A., et al.: Attention is all you need. In: Proceedings of Advances in Neural Information Processing Systems, pp. 5998–6008 (2017)
10. Kim, T., Yun, Y., Kim, N.: Deep learning-based knowledge graph generation for COVID-19. Sustainability **13**(4), 2276 (2021)
11. Yang, X., et al.: Clinical relation extraction using transformer-based models. arXiv preprint arXiv:2107.08957 (2021)
12. Qiao, B., Zou, Z., Huang, Y., Fang, K., Zhu, X., Chen, Y.: A joint model for entity and relation extraction based on BERT. Neural Comput. Appl. **34**, 3471–3481 (2021). https://doi.org/10.1007/s00521-021-05815-z
13. Pang, N., Qian, L., Lyu, W., Yang, J.D.: Transfer learning for scientific data chain extraction in small chemical corpus with joint BERT-CRF model. In: Proceedings of BIRNDL@ SIGIR, pp. 28–41 (2019)
14. Fader, A., Soderland, S., Etzioni, O.: Identifying relations for open information extraction. In: Proceedings of the 2011 Conference on Empirical Methods in Natural Language Processing, pp. 1535–1545 (2011)
15. Soares, L.B., FitzGerald, N., Ling, J., Kwiatkowski, T.: Matching the blanks: distributional similarity for relation learning. arXiv preprint arXiv:1906.03158 (2019)

Language-Free Regular Expression Search of Document's References

Aleksandr Ogaltsov[✉]

Antiplagiat Company, Moscow, Russia
ogaltsov@ap-team.ru

Abstract. In this paper, we address the problem of robust multilingual references section extraction not only from scientific papers, but a broader range of scientific documents. Although several publishers provide document's metadata including references separately, there are still many cases when it is extracted by hand. All existing methods of automation of this task work only with English language. It is motivated by the fact that international scientific literature is often written in English. However, there are many sources of scientific documents in national languages that are to be processed, for example, master or PhD theses. At the current time it is done mostly manually. We describe a simple yet efficient regular expression based method, which (1) takes as an input neither layout nor formatting features, but only plain text; (2) does not require any annotated multilingual data and performs in transfer learning manner; (3) achieves quality comparable to state-of-the-art multilingual BERT-based models with a substantial gain in terms of inference time.

Keywords: References extraction · Logical structure recovery · Multi-lingual metadata extraction

1 Introduction

The domain of intelligent document processing is currently growing, inspired by recent advances in the NLP community. The aim is to deeply understand documents to accelerate business processes by automatizing document workflow. Recognition of a document's logical structure is a crucial part of pipeline for many other downstream tasks such as language modeling, topic modeling and effective information retrieval. In this paper, we focus on a broad range of scientific documents and one special element of logical structure – the references section. There are many organizations such as online and offline libraries, scientific search engines and information aggregators that need to process document's metadata and references. Also, there are initiatives of cataloging references like MARC[1] and its national variants. Although in several cases publishers provide document's references and other metadata[2] separately, there are still many cases

---

[1] https://www.loc.gov/marc/unimarctomarc21.html.

[2] This research has done within the project of Russian governmental support of leading companies no. 1\549\2020 dated 19.06.2020.

E. Burnaev et al. (Eds.): AIST 2021, CCIS 1573, pp. 45–54, 2022.
https://doi.org/10.1007/978-3-031-15168-2_5

when it is extracted by hand, therefore automation of the process is needed. The second challenge is the multilingual perspective. The challenge arises because we consider not only papers that are massively written in English, but a broad range of a scientific documents like master or PhD thesis where national languages are dominating. Additional difficulty is a lack of consistency of bibliographic style across national scientific corpora. One of the frequent cases is when English is the main language but the bibliography can still contain references to national sources. Therefore we consider a multilingual setting for the task.

2 Related Work

References extraction task is not new. Much effort was dedicated to the problem in the early 2010s. Researchers used SVM [7], HMM [10], CRF-like [3] models, per line classification using Random Forest [2]. There are stable frameworks and tools for the problem [3,9,11] many of those are supported for now. There are two main drawbacks in this family of tools: (1) support of English language only, (2) many of those methods rely on layout or formatting features harvested from raw PDF [1]. Modern methods [8,12] can overcome the first drawback due to recent advances of multilingual text vectorization [4], [5], but still suffer from the second drawback; moreover, often modern methods consider document as an image [13,14]. There are several practical problems with layout or image consideration. First, some collections of documents initially do not have raw formatted versions but only plain text. Second, models trained with one formatting/layout extraction engine can hardly be portable to a new engine because the model is biased to the engine that was used as a feature extractor at training time. Such substitution naturally occurs when the engine updates, stops support, etc. Finally, formatting extraction and especially image conversion can take substantial time and make framework architecture more complex. On the other hand, plain-text extraction is fast and less affected by engine type. Moreover, small changes of plain-text should not affect bibliography extraction since it has a special text structure that can be recognized even in plain text. We propose a method for bibliography section detection that takes annotated plain text dataset in English and performs robustly in French, Russian, Chinese and Japanese without any tuning. The proposed method is based on simple regular expression search and described in Sect. 3, data and experiments are described in Sects. 4, 5 correspondingly.

3 Method

As we mentioned before, the bibliographic text has a special structure. Even if we have a plain text document in a foreign language on hands, we will rapidly distinguish the main body text from the bibliographic text. It often contains repeated initials, dates, page ranges etc. But how to cover all variants in all languages? Another question is what the minimal classification object: text line, sentence, token, etc.? Answers to these questions led us to the proposed method.

We developed eight regular expressions that do not rely on a particular language. We use \ p{L} for any letter in any language and preprocess text such that all international variants of a dot, comma and quotation marks (for example, ｡ ､ 「」) are mapped into single variant [.,"]. Actual used regular expressions are listed below:

1. `[\s;\.]\d{1,3}\.?\d?\s?\(?`
   `(\d{0,3}\-?\d{0,3}\)?)\s?[,:] \s?(\d{1,4}\s?\-\s?\d{0,4})`
   detects issue and volume, for example, 123 (45–67): 89
2. `[\s\-\.]\s?\p{L}{1,2}\s?\.\s? \d{1,4}[\s\.,]`
   detects page ranges, for example, p. 123
3. `[\s\-\.]\s?\p{L}{1,2}\s?\.\s? \d{1,4}\s?\-\s?\d{1,4}[\s;\.,]`
   detects page ranges, for example, pp. 123 - 456
4. `[\s\(\(\-](\p{L}{1,6}\.)\s?\.?`
   `\s?[,\.]?\s?(\d{1,4}\s? \(?\d{0,2}\)?[ivx])[;:,\.\s\)]`
   detects volume-like, for example, vol. 123
5. `\s(\d{1,3})\s?\(\s?(\d{1,3}\s? \-?\s?\d{0,3})\s?\)\s?[^\d ]`
   detects issue and volume, for example, 123 (45–67)
6. `\p{L}{1,14}[,\.]?\s\p{L}[,\.]? \p{L}?[,\.]`
   detects author abbreviation, for example, Ivanov A. B.
7. `\p{L}[,\.]\s?\p{L}?[,\.]?\s? \p{L}{1,14}`
   detects author abbreviation, for example, A. Ivanov
8. `\p{L}{1,14}[,]\s\p{L}\.`
   detects author abbreviation, for example, Ivanov, A.

The first five regular expressions are dedicated to finding different page ranges/volumes/journal numbers variants, and three others are used in order to find authors' initials. Note, that regular expressions are not designed to match *all* reference styles. They should match something in arbitrary place of bibliography. We hope that the concentration of matches is higher in references section than in other sections of the document. But since regular expressions are designed to be very common, they generate many false-positive detections in the main body text. The second step is to obtain the context of every detection and get simple language-independent features from it. In our case, there are eleven counters of different punctuation and digits in the detection context of 50 symbols from the left and from the right. If we have a dataset that is annotated with a position of bibliography in the text then each detection can be assigned with class label 1 if detection is located inside the bibliography and 0 otherwise. After that, we train logistic regression on obtained data. As a result, we have an algorithm to filter regular expression matches. We need high precision and some reasonable recall for positive class at this step to suppress most false-positive detections. The next step is to find the start and the end of the references section. We achieve this goal by analyzing the density of remaining matches after filtering. We search the position of bibliography beginning by keywords or by matches density – ranges with density above the mean are considered as bibliographic ranges. Figure 1 summarizes workflow of proposed method. Example of how suppression algorithm performs is presented in Figs. 2 and 3. The horizontal axis is the text symbol number, the vertical axis is the density of matches which is obtained by a pass of a sliding window of size 500 with a step size equals 1. Dashed horizontal line is mean density. Figure 2 shows matches density before suppression algorithm, Fig. 3 shows matches density after filtering by suppression algorithm with classification threshold equals to 0.6.

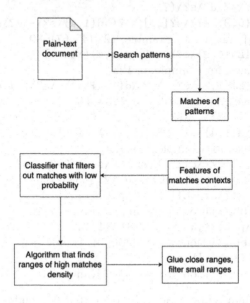

Fig. 1. Workflow of proposed method

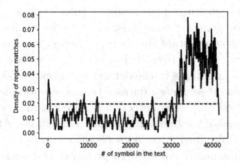

Fig. 2. Matches density before suppression algorithm. Many false-positive detections occur in the middle of the text

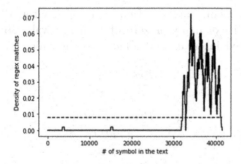

Fig. 3. Matches density after suppression algorithm. Density concentrates only around bibliography section

4   Data

We used synthetic data for the experiments. We considered five languages: English for train, development and test proposed method, French, Russian, Chinese and Japanese for test only without any tuning. We got 20K texts without bibliography section in English from Wikipedia[3]. The Bibliography section was generated with a random number of reference strings from a huge base of GIANT dataset [6] and pasted into every document with generated bibliographic keyword like "References" or without one. Each document contains 0 bibliographies in 10% cases, 1 bibliography in 80% cases and 2 bibliographies in 10% cases. If a document contains two reference sections, they are inserted in the middle of the document and the end of the document. In case of one bibliography, it is appended in the end of the document. The ground truth for each document are start-of-bibliography and end-of-bibliography symbols relative to the document text. Datasets for other languages generated by the same routine except bibliographies are sampled not from GIANT, but real documents downloaded from full-text access libraries like OATD[4] or institution sites. We downloaded documents by hand using language filter only. There was a small amount of documents to train on, so we kept a little part of real documents for test (see the end of Sect. 5). The other part was used as following. We copied bibliographies of those documents by hand and pasted them into the bibliographic base to sample from. We obtained approximately 500 annotated documents for each of French, Chinese, Japanese and 1000 for Russian[5].

5   Experiment

Quality measure We decided to choose the most common precision and recall scores relative to ground truth and predicted intervals with the format of

---

[3] https://en.wikipedia.org/.

[4] https://oatd.org.

[5] Generated data is available at https://bit.ly/bib_extraction.

<number of start symbol, number of end symbol>. Ground truth, as well as prediction, can consist of many or zero of such intervals. So, let *Pred* be the union of all predicted intervals and *Truth* be the union of all ground truth intervals, then

$$P = \frac{|Pred \cap Truth|}{|Pred|}$$

$$R = \frac{|Pred \cap Truth|}{|Truth|}$$

$$F1 = \frac{2PR}{(P+R)}$$

Proposed Method. First, we divided the dataset in English into train/dev/test sets in a proportion of 80/10/10 resulting in 16 K documents in train and 2 K in dev and test. Then we obtained 11 features for each regular expression match that was extracted from its context. Class label of the match is 1 if it was detected between start-of-bibliography and end-of-bibliography symbols and 0 otherwise. This leads to a training sample of approximately 8M objects. We trained logistic regression on this sample with the following quality in terms of binary precision and recall Table 1.

Table 1. Binary precision and recall for suppression algorithm

Label	Precision	Recall	Support
0	0.88	0.98	6.4M
1	0.87	0.49	1.7M

As we mentioned before, the positive class recall is enough to effectively suppress almost all matches in the body text, while leaving almost half of true matches. Next, we saved the start of every remaining match and calculate matches density by sliding window. After that, we tried to find bibliographic keywords for each language and consider them as section start. If we could not find keywords we considered as references section all ranges with match density above mean. Desired quality criteria are precision and recall relative to ground truth intervals and predicted intervals, where intervals are in the format (start, end) relative to text symbol number. Classification threshold and sliding window size were tuned on the development set by maximizing this quality criterion. A tuned algorithm was tested on the English test set and used for other languages without modifications.

LaBSE. To our knowledge there is no strong BERT-based baseline for the task, therefore we proposed one. We picked powerful BERT-based LaBSE architecture [5] that outputs multilingual vector representation of the sentence. For this part of the experiment, we slightly reformulated the task from interval prediction to text line classification. We split documents into lines and assigned each line with label 1 if it falls in bibliography interval and 0 otherwise. Then we obtained a 768-dimensional vector representation of each line as a features and trained logistic regression on top of it. For this experiment, we picked 5K documents from the original training set of 16K documents due to the very slow vectorization process. However, later on, it turned out that this fact didn't affect test quality. Results are listed in Table 2. One can see that fine-tuning of LaBSE with logistic regression gives almost perfect results for English.

Table 2. Binary precision and recall for logistic regression on top of LaBSE vector representation of each text line

Label	Precision	Recall	Support
0	1.00	1.00	260 K
1	0.99	1.00	37 K

In the test step, we converted algorithm predictions for each line into intervals relative to text for comparison in the same terms with the proposed method. We trained and developed this solution on the English dataset and tested if on the test part of the English dataset and other languages without modifications. Note, that LaBSE performance for English is almost perfect, however generalization to other languages is worse than proposed method.

ParsCit. We picked the off-the-shelf solution ParsCit [3], which solves the same problem of references section detection and can work with plain text. This is a pattern and CRF-based solution that was trained on English datasets. It outputs XML-markup of the initial document, so we converted it into text intervals and tested it by the proposed scheme, i.e. in a transfer-learning manner.

Results. See Table 5 for results of compared methods. Each cell indicates the F1 score defined in 5. Numbers in brackets for the proposed method correspond to the case when we completely turned off the bibliography start search by keywords. This is done in order to show that the proposed method does not completely rely on keyword search, however, keyword search is a good reinforcement that can gain 10–20 points of F1-score. Table 4 shows inference performances of the considered algorithms which were measured on the English test set. Performance of LaBSE algorithm was measured on Nvidia GeForce RTX 2080 Ti GPU, other algorithms on Intel Core i5 CPU. The proposed algorithm can process 30 documents per second, which is better than baselines by order of magnitude (Table 3).

We also tested proposed approach on small dataset of real documents. We obtained 39 documents in English, 36 in French, 27 in Russian, 15 in Chinese and 14 in Japanese. The document's type is mostly master, PhD and doctoral dissertations. These documents are different from the real documents that were used

Table 3. F1-scores for proposed method compared with LaBSE and ParsCit

Lang	Proposed	LaBSE	ParsCit
en (test)	0.86 (0.80)	**0.99**	0.95
fr	**0.86** (0.77)	0.82	0.82
ru	**0.87** (0.80)	0.73	0.86
zh	**0.82** (0.62)	0.74	0.45
ja	**0.84** (0.71)	0.70	0.29

Table 4. Inference performance of compared methods on the English test set

	Proposed (CPU)	LaBSE (GPU)	ParsCit (CPU)
docs/sec	**30**	1	2

Table 5. F1-scores for proposed method compared with LaBSE and ParsCit for real documents

Lang	Proposed	LaBSE	ParsCit
en	**0.9**	0.66	0.54
fr	**0.71**	0.55	0.18
ru	**0.85**	0.58	0.63
zh	**0.68**	0.51	0.59
ja	**0.64**	0.51	0.32

for synthetic dataset generation. Proposed model was trained fully on synthetic data in English.

One can see that the proposed method is shown to be quite robust across languages that were not presented in the training set at all. The combination of simplicity, interpretability and inference performance makes the proposed method a promising research direction.

Error Analysis. Here we present a brief error analysis of the proposed method. We divided errors into false positives (a text which is marked as a bibliography while it is not a bibliography) and false negatives (a text which is not marked as a bibliography, but should be marked). False negative detections occurred mostly in the bibliographies consisted fully of short references like "book name, year". False positive examples are mostly bibliographical footnotes and text around them. These parts of text were not labeled as a bibliographical blocks in ground truth, but still had high density of a regular expression matches.

6   Future Work

The main part of future work is to generalize the proposed method to the framework that would work in the proposed paradigm, but in a fully automated manner: given an annotated dataset, it should automatically generate regular expressions that will be common in terms of language and good in terms of discriminative power. After that, it will automatically construct a suppress algorithm and resulting interval search by density.

References

1. Ogaltsov, A.V., Bakhteev, O.Y.: Automatic metadata extraction from scientific pdf documents. Inform. Primen. **12**(2), 75–82 (2018)
2. Aleksandr Ogalstsov, K.S.: Automatic bibliography extraction from scientific papers. In: Mathematical Methods for Pattern Recognition. Russian Academy of Sciences, Moscow (2019)
3. Councill, I., Giles, C.L., Kan, M.Y.: ParsCit: an open-source CRF reference string parsing package. In: Proceedings of the Sixth International Conference on Language Resources and Evaluation (LREC'08). European Language Resources Association (ELRA), Marrakech, Morocco (2008)
4. Devlin, J., Chang, M., Lee, K., Toutanova, K.: BERT: pre-training of deep bidirectional transformers for language understanding. CoRR abs/1810.04805 (2018)
5. Feng, F., Yang, Y., Cer, D., Arivazhagan, N., Wang, W.: Language-agnostic BERT sentence embedding. CoRR abs/2007.01852 (2020)
6. Grennan, M., Schibel, M., Collins, A., Beel, J.: Giant: The 1-billion annotated synthetic bibliographic-reference-string dataset for deep citation parsing. In: 27th AIAI Irish Conference on Artificial Intelligence and Cognitive Science, pp. 101–112 (2019)
7. Han, H., Giles, C., Manavoglu, E., Zha, H., Zhang, Z., Fox, E.: Automatic document metadata extraction using support vector machines. In: Proceedings of 2003 Joint Conference on Digital Libraries, pp. 37–48 (2003). https://doi.org/10.1109/JCDL.2003.1204842
8. Krause, J., Shapiro, I., Saier, T., Färber, M.: Bootstrapping multilingual metadata extraction: a showcase in cyrillic. In: Proceedings of the Second Workshop on Scholarly Document Processing, pp. 66–72. Association for Computational Linguistics (2021). 10.18653/v1/2021.sdp-1.8, https://aclanthology.org/2021.sdp-1.8
9. Lopez, P.: GROBID: Combining automatic bibliographic data recognition and term extraction for scholarship publications. vol. 5714, pp. 473–474 (2009). https://doi.org/10.1007/978-3-642-04346-8_62
10. Ojokoh, B., Zhang, M., Tang, J.: A trigram hidden markov model for metadata extraction from heterogeneous references. Inf. Sci. **181**(9), 1538–1551 (2011)
11. Tkaczyk, D., Szostek, P., Fedoryszak, M., Dendek, P.J., Bolikowski, L.: CERMINE: Automatic extraction of structured metadata from scientific literature (2015)
12. Xu, Y., Li, M., Cui, L., Huang, S., Wei, F., Zhou, M.: LayoutLM: Pre-training of text and layout for document image understanding. In: Proceedings of the 26th ACM SIGKDD International Conference on Knowledge Discovery & Data Mining (2020). https://doi.org/10.1145/3394486.3403172, https://dx.doi.org/10.1145/3394486.3403172

13. Yang, X., Yümer, M.E., Asente, P., Kraley, M., Kifer, D., Giles, C.L.: Learning to extract semantic structure from documents using multimodal fully convolutional neural network. CoRR abs/1706.02337 (2017)
14. Zhong, X., Tang, J., Jimeno-Yepes, A.: PubLayNet: largest dataset ever for document layout analysis. CoRR abs/1908.07836 (2019)

Does BERT Look at Sentiment Lexicon?

Elena Razova[1] ⓘ, Sergey Vychegzhanin[1] ⓘ, and Evgeny Kotelnikov[1,2(✉)] ⓘ

[1] Vyatka State University, Kirov, Russia
kotelnikov.ev@gmail.com
[2] ITMO University, Saint-Petersburg, Russia

Abstract. The main approaches to sentiment analysis are rule-based methods and machine learning, in particular, deep neural network models with the Transformer architecture, including BERT. The performance of neural network models in the tasks of sentiment analysis is superior to the performance of rule-based methods. The reasons for this situation remain unclear due to the poor interpretability of deep neural network models. One of the main keys to understanding the fundamental differences between the two approaches is the analysis of how sentiment lexicon is taken into account in neural network models. To this end, we study the attention weights matrices of the Russian-language RuBERT model. We fine-tune RuBERT on sentiment text corpora and compare the distributions of attention weights for sentiment and neutral lexicons. It turns out that, on average, 3/4 of the heads of various model variants statistically pay more attention to the sentiment lexicon compared to the neutral one.

Keywords: Sentiment analysis · Sentiment lexicons · BERT · Interpretable models · Attention

1 Introduction

There are two main approaches to sentiment analysis of texts [3]: lexicon-based (or rule-based) and machine learning, in particular, using deep neural network models based on the Transformer architecture [31], including BERT [9]. The lexicon-based methods are high speed, easy to implement, they require no training data and a long learning process, and their results are easy to interpret [29]. However, neural network models achieve higher classification performance, for example, all state-of-the-art results for well-known sentiment analysis tasks (for example, SST, Yelp reviews, IMDB reviews, Amazon reviews) are achieved by neural network models.[1]

Despite much work on the study of deep neural network mechanisms [2, 32], the reasons for this success still remain incomprehensible. This is due to the high complexity of interpreting the neural network models containing hundreds of millions of parameters.

Taking the sentiment lexicon into consideration is one of the key aspects for understanding the differences between the lexicon-based approach and deep learning in sentiment analysis tasks. This paper explores the extent to which the BERT neural network

---

[1] https://paperswithcode.com/task/sentiment-analysis.

© The Author(s), under exclusive license to Springer Nature Switzerland AG 2022
E. Burnaev et al. (Eds.): AIST 2021, CCIS 1573, pp. 55–67, 2022.
https://doi.org/10.1007/978-3-031-15168-2_6

model pays attention to sentiment words. For this purpose, we construct the distribution of attention weights for the Russian-language neural network model RuBERT [18] for various subsets of words – a sentiment lexicon in general, positive and negative lexicons, as well as neutral one. Distributions are constructed for three variants of RuBERT models trained on three sentiment corpora (ROMIP 2012 News, SentiRuEval-2015 Banks and RuSentiment). We calculate and analyze distances between the constructed distributions based on Kullback-Leibler divergence. The Wilcoxon signed-rank test is used to test the significance of the distance discrepancy. The analysis made it possible to conclude that, on average, 3/4 of the heads of various variants of the RuBERT model statistically pay more attention to the sentiment lexicon compared to the neutral one.

The contribution of this article is as follows:

- we propose a method that analyzes how neural network models take into account various lexical subsets based on the distribution of attention weights and calculating the Kullback-Leibler distance between them;
- we analyze the accounting of sentiment and neutral words in the RuBERT model in the sentiment analysis task for three sentiment corpora;
- we conclude that on average RuBERT heads pay attention to sentiment words to a greater extent than to neutral ones, and this difference is statistically significant.

2 Previous Work

Interpretation of neural network models is carried out using three main approaches [2]: structural analysis, behavioral analysis and visualization.

Structural analysis examines various components of the neural network, such as word embeddings, sentence embeddings, attention weights, hidden layers, etc. [5, 36, 37]. The result of the analysis is to determine the role of the components in the neural network and to specify the type of information that these components can take into consideration.

Behavioral analysis consists in studying a neural network model on a set of test corpora, each of them reflects specific linguistic phenomena [1, 19]. Usually, most reference datasets are taken from text collections reflecting the natural frequency distribution of linguistic phenomena. Such sets are useful for assessing the average accuracy of a model, but may not reflect a wide range of linguistic phenomena. An alternative approach to assessing performance is to use different sets of tests, which can be created to investigate models for different tasks, natural languages, corpus sizes, etc.

Visualization is a complementary approach preceding structural and behavioral analysis [5, 8, 10]. Visualization helps to generate hypotheses about the behavior of a model or a dataset and to understand complex concepts.

Our research is carried out within the framework of structural analysis and the closest works are [5, 36].

Cao et al. [5] developed a differentiable masking method that allowed us to find out what different layers of the model "know" about the input data and where the prediction information is stored in different layers. The $BERT_{BASE}$ model was investigated for sentiment classification on the Stanford Sentiment Treebank (SST) [27]. Most layers have been found to rely on very positive or very negative words. In contrast to our work, the authors consider layers as a whole, without going down to the level of individual attention heads and specific weight distributions. They use annotated words in the SST as sentiment words, rather than a universal sentiment dictionary, as in our study.

Wu et al. [36] investigated the self-attention mechanism in the Transformer using the Layer-wise Attention Tracing method. SST and Stanford Emotional Narratives Dataset were used as corpora [24]. The authors have shown that attention weights have the highest values for very positive and very negative words. For this purpose, the dictionary by Warriner et al. was used, containing about 14,000 words with emotional valence [33]. The paper has also analyzed the proportion of attention given by individual heads to the sentiment words. In contrast to our work, [36] does not use BERT, but other Transformer-based models. They also use a sentiment dictionary, but to get the weight of the token, they sum up all the attention weights without considering the distribution of the weights. In our work, we build such distributions, which allows us to compare the weights for sentiment and neutral lexicons in more detail.

3   Interpretation Method

Our work aims to investigate the extent to which BERT-type models [9] pay attention to sentiment words. The research is based on the following hypothesis: BERT-type models pay different attention to sentiment and neutral lexicons. To test this hypothesis, the significance of the difference between the distributions of attention weights of senti-ment (positive and negative) words and neutral words has been evaluated. The distances between the distributions are calculated based on the Kullback-Leibler divergence.

The interpretation method involves four steps.

Step 1. Building an average attention matrix.

BERT uses WordPiece tokenization [35], whereby part of words is split into several tokens (subwords).[2] A word-level analysis of attention heads is required to calculate the attention that BERT assigns to sentiment words. Therefore, we convert the attention matrices of individual heads "token-token" into attention matrices "word-word". For this purpose, we implement the following procedure (see Fig. 1).

1. We summarize the attention weights directed to a tokenized word, by its tokens.
2. We take the average value of the attention weights over its tokens, directed from the split word to other tokens. This transformation preserves the property of the attention matrix, which means that the sum of the attention weights for each word is equal to one.
3. For each attention head and word, we find the average attention weight given to the word in the text.
4. We combine the separated words and lemmatize[3] all words of the text.

---

[2] RuBERT model for Russian uses BPE (Byte Pair Encoding) tokenization [26].

[3] The pymorphy2 [12] library was used for lemmatization.

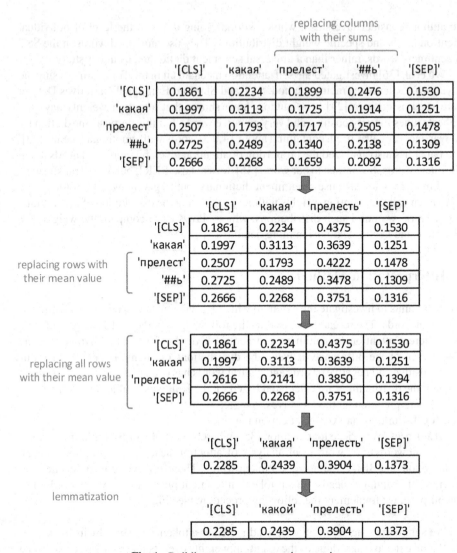

Fig. 1. Building an average attention matrix.

Thus, the first step results in finding the average attention weight given to this word in the text for each attention head and each word.

Step 2. Calculation of the distribution of attention weights for sentiment (W_s) and non-sentiment (neutral) words (W_{ns}) of the corpus. By non-sentiment words, we mean words that are not included in the sentiment dictionary. Sentiment words of the corpus are defined using the sentiment dictionary (see Subsect. 4.2).

To confirm the hypothesis that BERT pays different attention to sentiment and neutral lexicons, it is necessary to establish the difference in the distribution of attention weights of these types of lexicon. However, this difference can be accidental. To reduce the likelihood of a false conclusion, we ran 10 tests: on each test, we generated a random subset of neutral words for each attention head. The size of each random subset coincided with the number of sentiment words in the corpus.

We introduce the following notation:

- W_p – set of positive corpus words;
- W_n – set of negative corpus words;
- $W_s = W_p \cup W_n$ – set of sentiment corpus words;
- W_{ns} – set of non-sentiment (neutral) corpus words;
- $W_{e,i}$ – i-th random subset of neutral words of the set W_{ns}: $W_{e,i} \subset W_{ns}$, $|W_{e,i}| = |W_s|$, $i = [1, 10]$;
- $W_{o,i} = W_{ns}/W_{e,i}$ – set of neutral words after eliminating the i-th random subset.

At each of the ten tests, we determined the distribution of attention weights for words of five sets: sentiment (W_p, W_n, W_s) and neutral ($W_{e,i}$, $W_{o,i}$) words. The distributions of weights for all sets were calculated over one hundred value intervals (from 0 to 1 with a step of 0.01).

At the next step, the distances between the distributions of attention weights for sentiment (W_p, W_n, W_s) and neutral ($W_{o,i}$) words will be calculated, as well as the distances between the distributions of attention weights for subsets of neutral words $W_{e,i}$ and $W_{o,i}$. Comparing such distances enables to make a statistically significant conclusion about the difference in the distribution of attention between sentiment and neutral lexicons.

Step 3. Calculation of the Kullback-Leibler divergence between the distributions of attention weights of different word sets.

The distance between the distributions of attention weights was calculated based on the Kullback-Leibler divergence. The Kullback-Leibler divergence of the Q distribution relative to the P distribution (or "distance" from P to Q) is defined as follows [22, p. 34]:

$$D(P||Q) = \sum_{i=1}^{n} p_i log \frac{p_i}{q_i}. \tag{1}$$

For each set of neutral words $W_{o,i}$ and each h-th attention head, the following Kullback-Leibler distances are found:

$$\begin{aligned} D_{o,s,i}^h &= D(P_{W_{o,i}}^h || P_{W_s}^h), \\ D_{o,p,i}^h &= D(P_{W_{o,i}}^h || P_{W_p}^h), \\ D_{o,n,i}^h &= D(P_{W_{o,i}}^h || P_{W_n}^h), \\ D_{o,e,i}^h &= D(P_{W_{o,i}}^h || P_{W_e}^h), \end{aligned} \tag{2}$$

where $P_{W_{o,i}}^h$, $P_{W_s}^h$, $P_{W_p}^h$, $P_{W_n}^h$, $P_{W_e}^h$ – distribution of attention weights, respectively, of the sets of words $W_{o,i}$, W_s, W_p, W_n, W_e for the head h.

Next, for each attention head, the average Kullback-Leibler distances are found:

$$D_{o,s}^h = \frac{\sum_{i=1}^{10} D_{o,s,i}^h}{10},$$
$$D_{o,p}^h = \frac{\sum_{i=1}^{10} D_{o,p,i}^h}{10},$$
$$D_{o,n}^h = \frac{\sum_{i=1}^{10} D_{o,n,i}^h}{10},$$
$$D_{o,e}^h = \frac{\sum_{i=1}^{10} D_{o,e,i}^h}{10}. \tag{3}$$

Step 4. Testing the significance of differences between Kullback-Leibler distances for sentiment and neutral words.

To test the significance of the differences, we formulated two hypotheses:

- H_0 – the discrepancies between the Kullback-Leibler distances of sentiment and neutral words are random.
- H_1 – the discrepancies between the Kullback-Leibler distances of sentiment and neutral words are non-random.

To test these hypotheses, the Wilcoxon signed-rank test was used, which is designed to test the differences between two samples of independent measurements [34].

The Wilcoxon signed-rank test was applied at a significance level of $p = 0.05$ to evaluate hypotheses for the following pairs of Kullback-Leibler distances obtained in the previous step: $\langle D_{o,s}^h, D_{o,e}^h \rangle$, $\langle D_{o,p}^h, D_{o,e}^h \rangle$, $\langle D_{o,n}^h, D_{o,e}^h \rangle$.

Wilcoxon's test makes it possible to draw a conclusion only about the statistical significance of differences between samples of distances, but not about which distances prevail on average. For this purpose, we calculated the expected value of sets of distances.

Thus, the proposed interpretation method allows us to test and statistically substantiate the hypothesis that the BERT model pays different attention to sentiment and neutral lexicons, as well as to determine which lexicon is given more attention to.

4 Resources and Models

4.1 Text Corpora

In our experiments, we used three Russian-language corpora, annotated by sentiment: ROMIP 2012 News, SentiRuEval-2015 Banks and RuSentiment. We had two criteria when choosing corpora for analysis: 1) corpora should differ in the type of texts; 2) lexicon-based methods should show significantly lower performance for these corpora than neural network models.

The news corpus was prepared for the ROMIP 2012 sentiment analysis competition [7]. The corpus includes fragments of direct and indirect speech from news. The corpus of tweets about banks was prepared for the SentiRuEval-2015 sentiment analysis competition [21]. RuSentiment corpus includes posts on VKontakte [25]. We only used training parts of the corpora. The characteristics of the corpora are presented in Table 1.

Table 1. Characteristics of corpora, annotated by sentiment. The length of the texts is specified in tokens (mean ± std).[a]

Corpora	Total	Positive	Negative	Neutral	Length
ROMIP	4,260	1,115 (26%)	1,864 (44%)	1,281 (30%)	35 ± 28
SentiRuEval	4,883	354 (7%)	1,059 (22%)	3,470 (71%)	10 ± 5
RuSentiment	24,124	9,170 (38%)	3,654 (15%)	11,300 (47%)	13 ± 17

[a] Tokenization was carried out using NLTK (https://www.nltk.org).

Our experiments showed that for these corpora the difference between the results of the lexicon-based method (we used the version of the SO-CAL method [28] adapted for the Russian language) and the RuBERT model is the largest in comparison to other corpora. In particular, for ROMIP 2012 News, the difference in the macro F1-score for a three-class problem was 21% points (p.p.), for SentiRuEval-2015 Banks – 20 p.p., for RuSentiment – 27 p.p. [16].

4.2   Sentiment Dictionary

The sentiment dictionary must be highly accurate and complete. To create such a dictionary, we combined 9 publicly available Russian sentiment dictionaries [14, 16]: RuSentiLex [20], Word Map [17], SentiRusColl [15], EmoLex [23], LinisCrowd [11], Blinov's lexicon [4], Kotelnikov's lexicon [13], Chen-Skiena's lexicon [6], Tutubalina's lexicon [30]. The characteristics of the lexicons are shown in Table 2.

Table 2. The characteristics of sentiment lexicons.

Lexicon	Total	Positive elements		Negative elements	
		#	%	#	%
RuSentiLex	12,560	3,258	25.9%	9,302	74.1%
Word Map	11,237	4,491	40.0%	6,746	60.0%
SentiRusColl	6,538	3,981	60.9%	2,557	39.1%
EmoLex	4,600	1,982	43.1%	2,618	56.9%
LinisCrowd	3,986	1,126	28.2%	2,860	71.8%
Blinov's lexicon	3,524	1,611	45.7%	1,913	54.3%
Kotelnikov's lexicon	3,206	1,028	32.1%	2,178	67.9%
Chen-Skiena's lexicon	2,604	1,139	43.7%	1,465	56.3%
Tutubalina's lexicon	2,442	1,032	42.3%	1,410	57.7%

The final dictionary included only those words that are included in at least N source dictionaries. In accordance with our preliminary experiments, the optimal value for sentiment analysis based on the SO-CAL lexicon-based method is demonstrated by a dictionary with $N = 4$.

The final dictionary contains 2,313 words, including 823 positive (35.6%) and 1,490 negative (64.4%) words.

4.3   RuBERT Model

In our study the Russian-language neural network pre-trained model RuBERT [18] is used. The model was initialized on the basis of the multilingual version of $BERT_{BASE}$ and trained on the Russian part of Wikipedia and news articles.

The RuBERT model was fine-tuned (separately) on ROMIP 2012 News, SentiRuEval-2015 Banks and RuSentiment with the following parameters: number of epochs 5, batch size 8, learning rate 10^{-6}.

5   Results and Discussion

In the experimental part of the study, we applied the interpretation method proposed in Sect. 3 for three versions of the RuBERT model trained on three corpora: ROMIP 2012 News, SentiRuEval-2015 Banks and RuSentiment. All attention heads were examined only on the last (12th) layer.

Step 1. At the first step, we built the average attention matrices for each RuBERT variant.

Step 2. We calculated the distributions of attention weights of the sentiment (W_s) and neutral words (W_{ns}) of the corpus over one hundred value intervals (from 0 to 1 with a step of 0.01). An example of the distribution of attention weights for the seventh head of the last layer for the ROMIP 2012 News corpus is shown in Fig. 2.

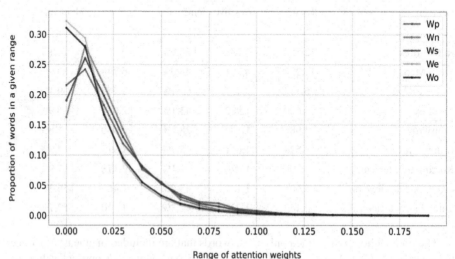

Fig. 2. Distribution of attention weights of sentiment and neutral words for the ROMIP 2012 News corpus for head #7 of the last layer.

Step 3. We calculated the Kullback-Leibler distances between the distributions of the attention weights of words.

Step 4. We tested the significance of the differences between Kullback-Leibler distances for sentiment and neutral words.

Table 3 shows the test results. Each cell contains two signs. The first sign is responsible for the result of checking the significance of the difference between the attention weights distributions of the corresponding sets of sentiment words ($sent - D_{o,s}^h$, $pos - D_{o,p}^h$, $neg - D_{o,n}^h$) and a set of neutral words $D_{o,e}^h$: "plus" indicates that the difference is significant, "minus" – not significant. The second sign is responsible for the ratio of the expected value of the corresponding distributions: "plus" means that the expected value of the attention weights distribution of the set of sentiment words is greater than the expected value of the distribution of the weights of the set of neutral words $D_{o,e}^h$; "minus" – less. At the bottom of the table, the number of "plus-plus" situations is given – the discrepancies are significant (non-random) and the expected value of the attention weights of the sentiment words is greater than the expected value of the attention weights of neutral words.

Table 3 shows that the difference of the attention weights distributions of sentiment words from the distributions of neutral words in 76.85% of cases is not random (the first sign in a pair is "plus"), and in 75% of cases, apart from the non-randomness of differences, the expected value of sentiment words is greater than the expected value of neutral words ("plus-plus" situation).

Table 3. Results of testing the significance of differences between Kullback-Leibler distances for sentiment and neutral words.

head	ROMIP 2012 News			SentiRuEval-2015 Banks			RuSentiment		
	sent	pos	neg	sent	pos	neg	sent	pos	neg
0	+ +	+ +	− +	+ −	+ −	− −	+ +	+ +	+ +
1	+ +	− +	+ +	+ +	− +	+ +	+ +	+ +	− +
2	− +	+ +	+ +	− +	+ +	+ +	+ +	+ +	+ +
3	+ +	+ +	+ +	+ +	+ +	+ +	+ +	+ +	− +
4	+ +	+ +	+ +	+ +	+ +	+ +	− +	+ +	− +
5	+ +	+ +	+ +	+ +	+ +	+ +	+ +	− +	+ +
6	− +	+ +	+ +	− +	+ +	+ +	+ +	+ +	+ +
7	+ +	+ +	+ +	+ +	+ +	+ +	+ +	+ +	+ +
8	− +	+ +	+ +	− +	+ +	+ +	+ +	+ +	− +
9	+ +	− +	+ +	+ +	− +	+ +	+ +	+ +	+ +
10	+ +	+ +	+ +	+ +	+ +	+ +	+ +	+ +	− +
11	+ +	− +	− +	+ +	− +	− +	+ +	− +	− +
+ +	9	9	10	8	8	10	11	10	6

The extent of attention for positive and negative lexicons, as well as for sentiment lexicon in general, does not differ notably. Table 3 shows that the picture is approximately the same for all three corpora (only for RuSentiment fewer heads pay attention to negative words than on average).

Interestingly, head #7 for all the corpora highlights sentiment words, both in general and positive/negative individually.

We have also noticed an interesting effect that is observed for all the three corpora: the expected value of the attention weights to neutral words decreases monotonically from the first head to the last.

We also looked at some examples of how attention weights to sentiment words affect the final decision of the model (Fig. 3).

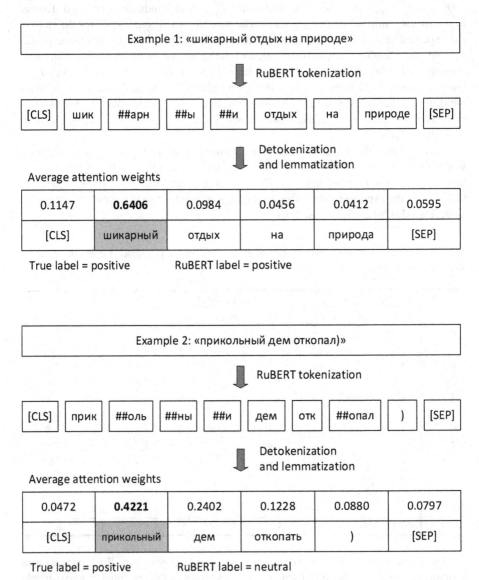

Fig. 3. Examples of texts and RuBERT attention weights.

In the first example (the RuSentiment corpus) RuBERT pays maximum attention to the sentiment positive word "шикарный" and the classification result coincides with the true label. However, in the second example (also from the RuSentiment corpus) RuBERT also pays maximum attention to the positive word "прикольный", but nevertheless decides on the neutrality of the text (with true label = "positive").

Thus, despite the high attention weights in relation to sentiment words, RuBERT does not necessarily make the final decision based only on them. Obviously, the decision-making process is more complex and requires further study.

6 Conclusion

Thus, we can conclude that 3/4 of heads, on average, pay more attention to sentiment words than neutral ones. The obtained results are consistent with studies [5, 36], but it is for the first time that we have obtained quantitative estimates of the degree of attention to sentiment words, verified on the basis of the statistical criterion.

In the future, we plan:

– to expand the sentiment dictionary – some sentiment words were not found due to insufficient size of the dictionary;
– to conduct a similar study on other corpora, annotated by sentiment;
– to improve the procedure for detokenization – some words after detokenization lost the letter "й" due to the peculiarities of the RuBERT tokenizer;
– to increase the number of tests from 10 to 100;
– to conduct an in-depth study of the attention weights, without averaging attention for words;
– examine attention on different layers.

References

1. Barnes, J., Ovrelid, L., Velldal, E.: Sentiment analysis is not solved! Assessing and probing sentiment classification. In: Proceedings of the ACL Workshop BlackboxNLP: Analyzing and Interpreting Neural Networks for NLP, pp. 12–23 (2019)
2. Belinkov, Y., Gehrmann, S., Pavlick, E.: Tutorial proposal: interpretability and analysis in neural NLP. In: Proceedings of the 58th Annual Meeting of the Association for Computational Linguistics, pp. 1–5 (2020)
3. Birjali, M., Kasri, M., Beni-Hssane, A.: A comprehensive survey on sentiment analysis: approaches, challenges and trends. Knowl.-Based Syst. **226**, 1–26 (2021)
4. Blinov, P.D., Klekovkina, M.V., Kotelnikov, E.V., Pestov, O.A.: Research of lexical approach and machine learning methods for sentiment analysis. In: Computational Linguistics and Intellectual Technologies: Proceedings of the International Conference "Dialogue", vol. 12, no. 19, pp. 51–61 (2013)
5. Cao, N.D., Schlichtkrull, M.S., Aziz, W., Titov, I.: How do decisions emerge across layers in neural models? Interpretation with differentiable masking. In: Proceedings of the Conference on Empirical Methods in Natural Language Processing, pp. 3243–3255 (2020)

6. Chen, Y., Skiena, S.: Building sentiment lexicons for all major languages. In: Proceedings of the 52nd Annual Meeting of the Association for Computational Linguistics, pp. 383–389 (2014)

7. Chetviorkin, I.I., Loukachevitch, N.V.: Sentiment analysis track at ROMIP 2012. In: Computational Linguistics and Intellectual Technologies: Proceedings of the International Conference "Dialog", vol. 2, pp. 40–50 (2013)

8. Clark, K., Khandelwal, U., Levy, O., Manning, C.D.: What does BERT look at? An analysis of BERT's attention. In: Proceedings of the ACL Workshop BlackboxNLP: Analyzing and Interpreting Neural Networks for NLP, pp. 276–286 (2019)

9. Devlin, J., Chang, M.-W., Lee, K., Toutanova, K.: BERT: pre-training of deep bidirectional transformers for language understanding. In: Proceedings of 7th Annual Conference of the North American Chapter of the Association for Computational Linguistics: Human Language Technologies (NAACL-HLT 2019), pp. 4171–4186 (2019)

10. Kim, S., Yi, J., Kim, E., Yoon, S.: Interpretation of NLP models through input marginalization. In: Proceedings of the Conference on Empirical Methods in Natural Language Processing, pp. 3154–3167 (2020)

11. Koltsova, O.Y., Alexeeva, S.V., Kolcov, S.N.: An opinion word lexicon and a training dataset for Russian sentiment analysis of social media. In: Computational Linguistics and Intellectual Technologies: Proceedings of the International Conference "Dialog", pp. 277–287 (2016)

12. Korobov, M.: Morphological analyzer and generator for Russian and Ukrainian languages. In: Khachay, M.Y., Konstantinova, N., Panchenko, A., Ignatov, D.I., Labunets, V.G. (eds.) AIST 2015. CCIS, vol. 542, pp. 320–332. Springer, Cham (2015). https://doi.org/10.1007/978-3-319-26123-2_31

13. Kotelnikov, E., Bushmeleva, N., Razova, E., Peskisheva, T., Pletneva, M.: Manually created sentiment lexicons: research and development. In: Computational Linguistics and Intellectual Technologies: Proceedings of the International Conference "Dialog", vol. 15(22), pp. 300–314 (2016)

14. Kotelnikov, E., Peskisheva, T., Kotelnikova, A., Razova, E.: A comparative study of publicly available Russian sentiment lexicons. In: Ustalov, D., Filchenkov, A., Pivovarova, L., Žižka, J. (eds.) AINL 2018. CCIS, vol. 930, pp. 139–151. Springer, Cham (2018). https://doi.org/10.1007/978-3-030-01204-5_14

15. Kotelnikova, A., Kotelnikov, E.: SentiRusColl: Russian collocation lexicon for sentiment analysis. In: Ustalov, D., Filchenkov, A., Pivovarova, L. (eds.) AINL 2019. CCIS, vol. 1119, pp. 18–32. Springer, Cham (2019). https://doi.org/10.1007/978-3-030-34518-1_2

16. Kotelnikova, A.V., Pashchenko, D.E., Kotelnikov, E.V., Bochenina, K.O.: Lexicon-based methods vs. BERT for text sentiment analysis. In: Proceedings of the 10th International Conference on Analysis of Images, Social Networks and Texts (AIST) (2021)

17. Kulagin, D.: Russian word sentiment polarity dictionary: a publicly available dataset. In: Artificial Intelligence and Natural Language. AINL 2019 (2019)

18. Kuratov, Y., Arkhipov, M.: Adaptation of deep bidirectional multilingual transformers for Russian language. In: Computational Linguistics and Intellectual Technologies: Proceedings of the International Conference "Dialog", pp. 333–340 (2019)

19. Lalor, J.P., Wu, H., Munkhdalai, T., Yu, H.: Understanding deep learning performance through an examination of test set difficulty: a psychometric case study. In: Proceedings of the Conference on Empirical Methods in Natural Language Processing, pp. 4711–4716 (2018)

20. Loukachevitch, N., Levchik, A.: Creating a general Russian sentiment lexicon. In: Proceedings of Language Resources and Evaluation Conference (LREC), pp. 1171–1176 (2016)

21. Loukashevitch, N.V., Blinov, P.D., Kotelnikov, E.V., Rubtsova, Y.V., Ivanov, V.V., Tutubalina, E.V.: SentiRuEval: testing object-oriented sentiment analysis systems in Russian. In: Computational Linguistics and Intellectual Technologies: Proceedings of the International Conference "Dialog", vol. 2, pp. 2–13 (2015)

22. MacKay, D.: Information Theory, Inference, and Learning Algorithms. Cambridge University Press, Cambridge (2003)
23. Mohammad, S.M., Turney, P.D.: Crowdsourcing a word-emotion association lexicon. Comput. Intell. **29**(3), 436–465 (2013)
24. Ong, D., Wu, Z., Tan, Z.-X., Reddan, M., Kahhale, I., et al.: Modeling emotion in complex stories: the Stanford Emotional Narratives Dataset. IEEE Trans. Affect. Comput. **12**, 570–594 (2021)
25. Rogers, A., Romanov, A., Rumshisky, A., Volkova, S., Gronas, M., Gribov, A.: RuSentiment: an enriched sentiment analysis dataset for social media in Russian. In: Proceedings of the 27th International Conference on Computational Linguistics, pp. 755–763 (2018)
26. Sennrich, R., Haddow, B., Birch, A.: Neural machine translation of rare words with subword units. In: Proceedings of the 54th Annual Meeting of the Association for Computational Linguistics (ACL), pp. 1715–1725 (2016)
27. Socher, R., et al.: Recursive deep models for semantic compositionality over a sentiment treebank. In: Proceedings of the Conference on Empirical Methods in Natural Language Processing, pp. 1631–1642 (2013)
28. Taboada, M., Brooke, J., Tofiloski, M., Voll, K., Stede, M.: Lexicon-based methods for sentiment analysis. Comput. Linguist. **37**(2), 267–307 (2011)
29. Taboada, M.: Sentiment analysis: an overview from linguistics. Ann. Rev. Linguist. **2**, 325–347 (2016)
30. Tutubalina, E.V.: Extraction and summarization methods for critical user reviews of a product. Ph.D. thesis, Kazan Federal University, Kazan, Russia (2016)
31. Vaswani, A., Shazeer, N., Parmar, N., Uszkoreit, J., Jones, L., et al.: Attention is all you need. In: Proceedings of the 31st Conference on Neural Information Processing Systems (NeurIPS), vol. 30, pp. 6000–6010 (2017)
32. Voita, E., Talbot, D., Moiseev, F., Sennrich, R., Titov, I.: Analyzing multi-head self-attention: specialized heads do the heavy lifting, the rest can be pruned. In: Proceedings of the 57th Annual Meeting of the Association for Computational Linguistics, pp. 5797–5808 (2019)
33. Warriner, A.B., Kuperman, V., Brysbaert, M.: Norms of valence, arousal, and dominance for 13,915 English lemmas. Behav. Res. Methods **45**(4), 1191–1207 (2013). https://doi.org/10.3758/s13428-012-0314-x
34. Wilcoxon, F.: Individual comparisons by ranking methods. Biometrics Bull. **6**(1), 80–83 (1945)
35. Wu, Y., Schuster, M., Chen, Z., Le, Q.V., Norouzi, M., et al.: Google's neural machine translation system: bridging the gap between human and machine translation. arXiv:1609.08144 (2016)
36. Wu, Z., Nguyen, T.-S., Ong, D.: Structured self-attention weights encode semantics in sentiment analysis. In: Proceedings of the Third BlackboxNLP Workshop on Analyzing and Interpreting Neural Networks for NLP, pp. 255–264 (2020)
37. Wu, Z., Ong, D.C.: On explaining your explanations of BERT: an empirical study with sequence classification. arXiv:2101.00196 (2021)

Context-Based Text-Graph Embeddings in Word-Sense Induction Tasks

Leonid Sherstyuk[1,3]($\boxtimes$) and Ilya Makarov[2,4] (iD)

[1] HSE University, Moscow, Russia
lsherstyuk@gmail.com
[2] Artificial Intelligence Research Institute (AIRI), Moscow, Russia
iamakarov@hse.ru
[3] Sber, Moscow, Russia
[4] Big Data Research Center, National University of Science and Technology MISIS, Moscow, Russia

Abstract. Word Sense Induction (WSI) is the process of automatically discovering multiple senses or meanings of a word. WSI task can be described as grouping contexts of a given word by its senses which are not provided beforehand. Modern WSI systems are given small text fragments only and should cluster them into some unidentified number of clusters. In the present work contextualized word embeddings, calculated by BERT, are applied in conjunction with clustering techniques to the WSI task for the Russian language. We hypothesize that novel language model embeddings, already viable for sense induction, may be enhanced by graph-based post-processing. We evaluate that proposition on 3 datasets from the Russian language WSI competition task. Fusion of graph algorithms and vector representations allowed us to beat one of the tasks' baseline (wiki-wiki, ARI = 0.7513) and demonstrate viability of further research. This work provides insight into how vector sentence representations can be organized for more efficient sense extraction.

Keywords: NLP · Word Sense Induction · BERT · Graph clustering

1 Introduction

In Natural Language Processing (NLP), the ambiguity of words in sentences has long been one of the most compelling challenges. Two tasks were formulated to address this problem: Word Sense Disambiguation (WSD) and Word Sense Induction (WSI).

Word Sense Disambiguation task is concerned with selecting the correct sense of the word out of some given group. Consider the following sentences: *Mary went to the bank for a loan* and *The bank of the river was covered with snails*. We would be solving the WSD task, if, presented with these sentences, we were asked to tell, where the word *bank* is describing financial institution, and where the *bank* describes land patch by the water body. The set of these senses would be called sense inventory, and be available beforehand. In the WSI setting, on the other

E. Burnaev et al. (Eds.): AIST 2021, CCIS 1573, pp. 68–81, 2022.
https://doi.org/10.1007/978-3-031-15168-2_7

hand, we are only asked to distinguish one sense from the other. WSI tasks are also commonly called unsupervised WSD, as no target labeling is given. Thus, WSD usually requires massive structured databases of definitions and senses, while WSI enables knowledge-free approaches to sense discovery.

The reason for the complexity of these tasks lies in the inherent ambivalence of natural text; something obvious for a human reader may be tricky to formalize automatically, and the sense definitions are usually fuzzy. Meanwhile, advances in solving the task would be fruitful for many user-facing applications, such as machine translation and chatbots.

Recent achievements in context-based language modeling, such as the creation of ELMo [25] and Transformer [32], extend the reach of adjacent tasks. Vector representations of text, created by models of that kind, are often enough by themselves for solving a wide variety of tasks.

In this work we will attempt to solve the WSI task, utilizing contemporary language models and applying dimensionality reduction techniques to optimize for potential product performance. We will assess several parameters of preprocessing text embeddings and test several classical graph clustering approaches.

We will evaluate our approaches on tasks of RUSSE'2018 - competition for NLP on Russian language data.

2   Related Work

In the following section, we will review the first approaches to these questions and introduce contemporary research on the topic.

Knowledge-Based Approaches. The development of automated sense disambiguation and induction has been made possible by formulating distributional hypothesis [10]; it posits that word, used in a specific sense, tends to be in the same context, leading to the assumption that context word set may be enough for sense encoding.

Early approaches to the WSI problems featured manually composed frames, describing fine-grained senses of the word [27], having discovery of sense being done by dictionary lookup. [13] enhanced this approach by comparing dictionary definitions of senses and sentence and considering word overlap as a measure of similarity and confidence; this technique works, as long as the sense definition and wording in the original sentence are comprehensive, which is rarely true. Another way to improve on the approach is to create more encompassing system for senses storage, which incorporates semantic structure of language, conveyed by characteristics, such as grammatical and morphosyntactic features of words, as well as their rich semantic neighborhood; most notable attempts include Wordnet [21] and BabelNet [22].

Graph-Based Approaches. The introduction of lexical graph databases greatly empowered graph-based approaches to WSI tasks. For instance, classical graph algorithms like PageRank and HITS were shown to be successful and sufficiently fast for the task [2]. A more ubiquitous approach, though, is to discover senses through graph clustering, which also performs well [11]. [31] introduced meta-clustering algorithm, which follows hard-clustering of context by fuzzy sense clustering. We have also successfully applied graph [16–18] and text [15,19,26,29, 30] based embeddings in our previous studies for downstream machine learning problems.

Representing a corpus as a graph enables researching more complicated data structures. [3] induces sense for tags in internet communities and points out that, rather than working with tags and text as separate entities, triplets of tag-resource-user performs much better. [8] suggests the applicability of social network analysis methods in conjunction with clustering and knowledge trees to effectively induce senses in internet corpora.

Further developments of graph-based approaches, such as enrichment of graph information with additional features, e.g. syntactic characteristics of the sentence [9], don't solve its core issues with the particularity of the sense definitions and immutability of semantic connections, which doesn't go well with inherently fuzzy senses of words. This problem calls for new semantic representation models.

Distributional Context Embedding Models. [20] introduced word2vec models - feed-forward neural network algorithms, which may be trained to either predict a word given its context (Skip-Gram) or vice versa (CBoW). Resulting embedding into vector space tends to be effective in capturing semantic content of a word, and its addition to already present processing pipelines gives a substantial quality boost, i.e. [5]. Additionally, CBoW-like models allow for knowledge-free approaches to emerge: processing a large corpus of text enables unsupervised word sense discovery through calculation and clustering of distributional representation of the words, i.e. [24].

The key issue of the word2vec approach - independence of words in the context - is first solved with ELMo [25] by introducing deeper inter-dependence of the target word and its context through means of bi-directional LSTM model, encoding a word's meaning, rather than a word itself.

[32] introduces the Transformer - an encoder-decoder language model, which performs better than ELMo on various tasks, but fails to capture both the left and right context of the word. BERT [7] resolves this problem by introducing word masking during the training. The unsupervised nature of BERT's training loop makes possible its training on a large corpus of raw text and enables transfer learning.

[12] introduces several pre-trained BERT models, trained on Russian resources:

- RuBERT is trained Wikipedia and literature corpora, and has 180M parameters - that's 1.5 times the size of the BERT-base model, the most ubiquitous iteration of BERT. It is considered to be a state-of-the-art BERT-like Russian language model available at the time of writing this work.
- Conversational RuBERT was trained on corpora, compiled from Russian entertainment social networks, as well as subtitles database. It was initialized by RuBERT and roughly equal to it in terms of the number of parameters.

WSI in Russian. The first big collection of evaluation datasets on the Russian language has been released during the RUSSE competition in 2018 [23]. Organizers provided 3 datasets with sense labels. No sense inventory was provided, so participants were free to extract senses' content for any resource they want.

Pre-trained CBoW turned out to be the most successful model of the competition: test data was embedded and followed by the calculation: most frequent word(s) - 1st sense, frequent but distant (in the embedding space) word - second sense, etc.Other models include:

- Affinity propagation on CBoW embeddings, discovering weighted average of the words in the context
- Ward clustering on normalized sums of fastText embeddings (pretrained on lib.rus.ec).
- Agglomerative clustering on a weighted average of word2vec vectors for context.
- Neural network with self-attention to encode sentences, trained on wiki, librusec, and train dataset, followed by k-means.

Later works applied BERT architecture for solving Russian WSI tasks. [28] implements and trains Transformer models, variating number of layers (1 to 2), heads (1 to 4), addition of attention and tf-idf weighting; they get comparable results and beat some of the best ones.

In this work we will use a similar model - RuBERT, pre-trained on a large corpus of texts and described in the previous section. What we will explore is how graph clustering methods are performing, when being used in conjunction with semantically rich BERT embeddings.

3   Methods and Data

3.1   Data

RUSSE competition provides three main datasets, split into train and test subsamples. Instead of sense inventory, gold-standard labels are provided, which are different for different senses. Frequency characteristics of datasets are present in Table 1. In the following sections, we will briefly review these datasets.

Table 1. Dataset statistics in the shared task.

Dataset	Inventory	Corpus	Split	# of words	# of senses	Avg. # of senses	# of contexts
wiki-wiki	Wikipedia	Wikipedia	Train	4	8	2.0	439
wiki-wiki	Wikipedia	Wikipedia	Test	5	12	2.4	539
bts-rnc	BTS	RNC	Train	30	96	3.2	3491
bts-rnc	BTS	RNC	Test	51	153	3.0	6556
active-dict	Active Dict	Active Dict	Train	85	312	3.7	2073
active-dict	Active Dict	Active Dict	Test	168	555	3.3	3729
active-rnc	Active Dict	RNC	Train	20	71	3.6	1829
active-rutenten	Active Dict	ruTenTen	Train	21	71	3.4	3671
bts-rutenten	BTS	ruTenTen	Train	11	25	2.3	956

Wiki-wiki dataset was constructed from homonyms, which have separate articles for each sense in the Russian Wikipedia. The assumption was that, given a Wikipedia article containing an ambiguous word in its title, all its occurrences throughout the article will share the same sense. Parsed contexts were checked for sufficiency in word count and reviewed by organizers; 9 homonyms were left.

Word set of the BTS-RNC dataset is based on the Russian Large Explanatory Dictionary [1] (Bolshoy Tolkovyj Slovar', BTS), and contexts were samples from the Russian National Corpora (RNC). Two kinds of polysemy were used: polysemy with metaphorical senses and homonymy. Out of 30 train words, 21 are homonyms; out of 51 test words, 40 are homonyms. Senses were described by linguistics specialists and labeled through crowdsourcing.

Active-dict dataset [14] was constructed from the Active Dictionary of Russian - an explanatory dictionary, based on modern language. Word senses are constructed manually and are considered distinct if their semantic and syntactic characteristics differ. Train and test have 85 and 168 words respectively, each containing only one homonymous word.

Evaluation. We will use Adjusted Rand Index, which was also used during the competition, which solves many problems, specific to clustering evaluation (no clear worst-case scenario, corner cases like "one big cluster" or "one iterm - one cluster", etc.). The Rand Index computes a similarity measure between two clusterings by considering all pairs of samples and counting pairs that are assigned in the same or different clusters in the predicted and true clusterings. The raw RI score is then "adjusted for chance" into the ARI score using the following formula: $ARI = \frac{(RI - E[RI])}{(max(RI) - E[RI])}$

3.2   Methods

As we have described above, the goal of this work is to explore the applicability and efficacy of graph clustering algorithms, when used in conjunction with BERT embeddings. Broadly, our WSI pipeline has the following steps:

1. Computation of BERT embedding for the target contexts
2. Selection and preprocessing of the embeddings
3. Construction of context similarity matrix, construction, and clustering of the graph

We will describe each step of our pipeline in the sections below.

Language Models. We use RuBERT and Conversational RuBERT to create word and context embeddings. Pre-trained BERT-like models are sufficiently parameterized to practically eliminate the need for additional fine-tuning.

Conversational RuBERT embeddings are expected to perform better on the *active-dict* dataset, as it is comprised overwhelmingly of metaphorical polysemy words, which is more characteristic of the speech, than of literature-like writing.

Throughout this work we use notion of "embeddings" to refer to the content of the last hidden layer of the BERT model.

Embedding Preprocessing. In this section, we will review parameters, which were studied during the embedding preprocessing. Their list is the following:

- Selection of embeddings
- Similarity metric
- Vector normalization
- Reduction of the number of connections
- Binarizing of similarities

The first parameter to survey is the selection of embeddings. One strong benefit of using BERT embeddings is that it returns vectors for each word, but each vector is highly enriched with semantic context information from the rest of the sentence. For the next step, we will require to construct a similarity matrix between contexts. The following configurations were considered:

1. similarity between embeddings of the target words;
2. similarity between pooled embedding of the context;
3. minimal similarity between any pair of word embeddings for any two contexts;
4. minimal similarity between any pair of word embeddings for any two contexts shifted towards the target word of context;

Set of word embeddings becomes a cluster itself, and its proximity to other target word's senses representation may not always be best expressed with the target word embedding. We denote the closest neighbor distance of these clusters as minimal distance, used to find minimal similarity.

(1) takes embedding of the target word. (2) uses embedding of the first technical token - it is said to capture the context of the whole sentence. (3) takes embeddings of every word in sentence and performs minimal linkage between contexts. (4) is similar to the previous mode, but every embedding is averaged with target word's embedding; that way, we assume, connections between contexts would be more likely conditioned by the target word semantics.

As a similarity measure we employed three distances, which are later transformed into similarities: Cosine, Manhattan and Euclidean distances.

We had an option to apply the L2 vector normalization to the embeddings vectors before similarity calculation.

We also experiment with the reduction of the number of connections. By performing the aforementioned preprocessing, we get a pairwise similarity matrix of contexts. For graph methods to be more effective, we need to downgrade weak ties, leaving only the strong ones. We did that by filtering out all edges, leaving only top-K row-wise values, setting all the rest to zero.

The last option the binarization of the similarity matrix, setting to 1 all non-zero values. We hypothesise that truncating information on the degree of similarity to the boolean "connected-disconnected" ties would be beneficial to the inference.

Graph Construction and Clustering. The resulting similarity matrix is used as an adjacency matrix to construct a contexts graph. We used three graph clustering algorithms, which were shown to be effective in the literature:

- Chinese Whispers (CW)
- MaxMax (MM)
- Label Propagation (LP)

MM and LP are parameter-free algorithms, which makes them much more favorable for training and inference settings. CW [4] has only one parameter.

4 Results

In this section, we will review the impact of every individual parameter, and assess the overall performance of the algorithms.

4.1 Parameter-Wise Performance

Each following subsection reviews performance for each parameter group. All parameter-wise results may be found in Table 2. Each section was compiled with the following procedure: we selected the best ARI score for each dataset, where the parameter in question takes the value, provided in a row label. That way we can ascertain the upper boundary of the performance with each parameter value. Table 3 provides best parametrization for every dataset.

Language Model. Our experiments included two language models: RuBERT, trained on Wikipedia and literature corpus, and Conversational RuBERT, trained on social media and entertainment social networks datasets. Conversational RuBERT's top performance is better or on par with the RuBERT. That is surprising, as RuBERT performed worse even on the *Wiki-wiki* dataset, though Wikipedia was a big part of its training sample.

Embedding Selection. Our experiments covered 4 modes of embedding selection and adjacency table construction. Two of the approaches - *Target word* and *Pooled context* just passed embeddings, produced by BERT, further down the pipeline. *Minimal context* approach calculated minimum neighboring pairwise distance between words of every context. *Minimal context with target shift* calculated the same distance, but every vector was transformed into its average with the corresponding target word's embedding.

Pooled context performed the worst on every dataset, which is not surprising - whole semantic information of the sentence was condensed into one vector, while other approaches didn't reduce dimensionality and task relatedness as dramatically. Target word is ranked second for BTS-RNC and Wiki-wiki datasets and third for Active-dict. This is surprising, as it could be thought of as the most straightforward way to approach our task. Minimal context shown the best performance on Active-dict but practically failed on the other two datasets.

Table 2. Best results for each parameter group; best results in parameter group per dataset in bold

Variable	Value	Dataset		
		Active-dict	BTS-RNC	Wiki-wiki
BERT model	Conversational BERT	0.140	**0.195**	**0.751**
	Classic BERT	0.140	0.186	0.678
Embedding selection and processing	Minimal context	**0.140**	**0.139**	0.521
	Minimal context w/ target shift	0.136	0.195	**0.751**
	Pooled context	0.035	0.013	0.444
	Target word	0.124	0.186	0.657
Distance	Cosine	0.129	**0.195**	**0.751**
	Euclidean	**0.140**	0.178	0.659
	Manhattan	0.137	0.172	0.678
Vector normalization	No normalization	0.129	**0.195**	**0.751**
	L2	**0.140**	0.186	0.665
Number of connections	2	0.140	0.154	0.341
	5	0.140	0.174	0.455
	7	0.140	0.193	0.634
	10	0.140	**0.195**	**0.751**
Binarization of similarities	Raw	**0.140**	0.188	0.665
	Binarized	0.136	**0.195**	**0.751**
Clustering algorithm	ChineseWhispers	0.140	0.193	**0.751**
	LabelProp	0.140	**0.195**	0.659
	MaxMax	0.140	0.118	0.593

Table 3. Best model parametrization for each dataset; "–" denotes same results for each parameter value

	Active-dict dataset	BTS-RNC dataset	Wiki-Wiki dataset
Kind of BERT	–	ConvBERT	ConvBERT
Embedding selection and processing	Minimal context	Minimal context with target shift	Minimal context with target shift
Distance	Euclidean	Cosine	Cosine
Vector normalization	L2	No normalization	No normalization
Number of connections	–	10	10
Binarization of similarities	Raw	Binarized	Binarized
Clustering algorithm	MaxMax	LabelProp	ChineseWhispers

Minimal context with target shift shows the best performance on Wiki-wiki and BTS-RNC, slightly underperforming on Active-dict. We can interpret it that our intuition was correct: the set of word embeddings is more useful in connecting with other contexts than one word, context vector, and target shift merely emphasizes our interest in the sense of the exact target word.

Similarity metrics didn't amount to marginal performance improvement on any dataset, but cosine distance was the best for the two datasets. Euclidean distance performed worst on Wiki-wiki and best on Active-dict, while Manhattan distance's performance was average. It makes sense to rely on Cosine metrics most of the time, and have Euclidean distance as a backup distance function.

Vector normalization didn't increase performance too much - except for Active-dict, but it didn't hinder the scores either. To evaluate reduction of the number of connections we select the top-K closest contexts for each context; 4 values of K were tested.

Connection number did not affect the Active-dict dataset, while other datasets were clustered the better the higher value of K came. It is possible we have not reached an optimal value of K, which is why the highest value of the parameter has the highest score.

Binarization of similarity matrices was performed to test whether the actual degrees of similarity contain useful information in terms of clustering, or just having boolean logic (edge or no edge) is enough. Binarization seems to be beneficial on the BTS-RNC and Wiki-wiki datasets.

Clustering algorithms - three clustering algorithms were considered. Two of them - ChineseWhispers and LabelProp have been long used for solving WSI tasks, and MaxMax was proposed by its authors as the soft-clustering algorithm to be used in WSI tasks. Performance on Active-dict isn't impacted by the choice of clustering. CW and LP both performed well on BTS-RNC, while MM produced substantially worse results. The clear leader on the Wiki-wiki dataset is CW. CW performed best on all three datasets, while LP produced comparable or worse results. MM displayed the least successful performance.

4.2   Overall Performance

As was shown above, the usefulness of particular parameters is very sensitive to the evaluation dataset. The Table 4 summarizes the importance of the factors for each dataset in the following way: out of the best models for each value of a parameter, we compute the difference between the highest and lowest values. That way we can evaluate sensitivity for each parameter-dataset pair.

Embedding selection displays the biggest discrepancy for all three datasets, followed by the choice of the clustering algorithm. Another notable difference in metrics is present in neighbours count for wiki-wiki dataset.

5   Discussion

In this section, we will attempt to interpret the highlights of the experiments, and put our final scores into the perspective of prior work.

Table 4. Maximum difference between best models, which are different in the pipeline parameter; most notable discrepancies in bold

Hyper-parameter	Active-dict	BTS-RNC	Wiki-Wiki
Type of BERT	0	0.008	0.072
Clustering algorithm	8.9E-05	**0.077**	**0.157**
Distance function	0.010	0.022	0.091
Embedding selection	**0.104**	**0.181**	**0.306**
Similarities binarization	0.003	0.006	0.085
Embed normalization	0.010	0.009	0.085
K in KNN reduction	0	0.041	**0.409**

Embedding selection is by far the most important factor of the algorithm. Our initial idea was that target word embedding was sufficient for determining a sense of the word: it is the most relevant to the actual word, sense of which we're trying to induce, and it should contain sufficient information about the rest of the sentence for it to be connected to the same-sense words. Our research suggests that individual words from the context play a key role in successful sense inducing. We used minimal-context idea, and, when in conjunction with the target word, it greatly outperforms other approaches. That idea is reminiscent of the classical approach in WSI - co-occurrence graphs. It could be improved even further by identifying all of the words in the sentence, relevant to the target and its sense (we identified only the target word itself) and using them to calculate the similarity between contexts.

Clustering algorithm is the next biggest impact factor. We consciously selected parameter-free algorithms to speed up the inference time, which may be suboptimal. The most unfortunate result is the underperformance of the MaxMax algorithm, which may be due to our preprocessing being too aggressive in reducing information about the senses, so the algorithm is struggling to provide good results.

Similarity graph reduction, performed by selection of K nearest neighbors, has provided the biggest discrepancy value of all. it is clear that, rather than performing too good with optimal values, it is the case that algorithm performance is too weak due to bad results being too bad. Aggressive edge pruning seems to be reducing score dramatically in every setting, so it is to be avoided. Computation over a fully connected graph, while being expensive resource-wise, is absolutely is up for exploring. Another avenue for improvement is to set K as a fraction of all neighbors, rather than imposing absolute value. That way, its performance would be more stable for the corpora, where the number of contexts varies too much.

BERT model we're using is on the lower end of influencing the results. In retrospect it is easy to explain the superior performance of Conversational RuBERT:

it was initialized by a base RuBERT, and, consequently, exposed to more of the training samples. Though their parameter count is roughly the same, ConvRuBERT was exposed to more training, and therefore, tends to have a richer language model.

Other parameters, such as embedding normalization, similarity binarization, and distance function are hard to discuss, as the discrepancy they produced wasn't too great, and their effectiveness is unclear, as the dataset itself had much more impact than the parameters did. We will discuss questions, raised by the data consistency, in the following section.

Table 5. Best scores comparison, best per dataset in bold

	Active-dict	BTS-RNC	Wiki-wiki
RUSSE best	0.247	**0.338**	**0.962**
RUSSE baseline	0.153	0.213	0.527
Davydova best	0.196	0.303	0.789
Struyanskiy & Arefyev best	**0.306**	0.316	0.651
Out best models	0.140	0.195	0.751

Benchmarking. Finally, we shall compare our results with the analogous work in the field. We chose best and baseline results from the competition, as well as two papers on WSI, which were evaluated on the same dataset:

- [28] considered various configurations of Word2Vec models, TF-IDF, as well as self-trained BERT-like architecture.
- [6] explored pre-trained BERT and ELMo models with UMAP clustering on the embeddings of the target word.

Results of the models are presented in the Table 5. In two cases out of three, we came very close to the competition baseline, exceeding it on one dataset.

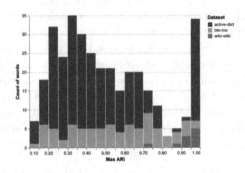

Fig. 1. Distribution of max ARI scores per word per dataset

Datasets. Both *Active-dict* and *BTS-RNC* datasets are problematic for every researcher - max ARI score doesn't exceed 0.35. On the other hand, *Wiki-Wiki* is almost completely solvable back at the competition. As both problematic datasets were constructed manually by linguistics experts, it may be suggested that dataset compilation methodology may be the culprit. Conversely, the *wiki-wiki* dataset was parsed from a very ubiquitous resource, which is even part of the many training samples for many contextual models. It is unclear if the problematic datasets are too hard, or the wiki dataset is too easy, but the discrepancy in their max scores indicates the need for a thorough assessment, as robust datasets are an important part of pushing the research field further.

To further explore this, we can look at the distribution of the max scores per word (see Fig. 1). Most of the easiest words to induce are coming from active-dict dataset, but the rest of the scores are skewed to the left. The dataset is highly imbalanced - senses are either too easy to predict or too hard. In contrast, *BTS-RNC*'s scores are distributed uniformly.

6   Conclusion

In this work we explored solving word sense induction in Russian language task with a mix of contemporary language modeling via BERT and more classic graph clustering methods. Our experiments were focused on exploring the impact of preprocessing parameters and algorithm selection. We evaluated the impact of seven parameters, two language models, and three clustering techniques. Our results are on par with the competition baselines, exceeding them on one out of three cases. We discovered an effective heuristic for embedding selection and processing, which is based on an idea of using co-occurrence graphs but applies to vector embeddings of BERT models, demonstrating the relevance of the classic graph approaches when used in conjunction with contemporary language models.

Our empirical results delineate the ways for the work to be advanced. Our discovery of minimalization of context similarity may be extended, borrowing more ideas from the practice of the use of co-occurrence graphs. Also, parametrized clustering techniques may be explored for performance improvement. Automated data-dependent hyperparameter extraction - for instance, nearest neighbors number - may greatly enhance final results.

Acknowledgement. This work was financially supported by the Ministry of Science and Higher Education of the Russian Federation, contract 075-11-2022-36.

References

1. KS, A.: Large Explanatory Dictionary of Russian [Bol'shoy Tolkoviy Slovar' Russkogo Yazika]. Norint (2014)
2. Agirre, E., et al.: Two graph-based algorithms for state-of-the-art WSD. In: Proceedings of the CEMNLP 2006, pp. 585–593. ACL, USA (2006)
3. Andrews, P., Pane, J.: Sense induction in folksonomies: a review. Artif. Intell. Rev. **40**(2), 147–174 (2013)

4. Biemann, C.: Chinese whispers - an efficient graph clustering algorithm and its application to natural language processing problems. In: Proceedings of TG, pp. 73–80. ACL, New York City (2006)
5. Chang, H.S., et al.: Efficient graph-based word sense induction by distributional inclusion vector embeddings. arXiv preprint arXiv:1804.03257 (2018)
6. Davydova, V.: Context-dependent word embeddings for word sense induction for Russian language (master thesis) (2020)
7. Devlin, J., Chang, M.W., Lee, K., Toutanova, K.: BERT: pre-training of deep bidirectional transformers for language understanding. In: Proceedings of the ACL 2019, pp. 4171–4186. ACL, Minneapolis (2019)
8. García-Silva, A., et al.: Discovering and associating semantics to tags in folksonomies. Knowl. Eng. Rev., 1–24 (2004)
9. Goularte, F.B., et al.: Msc+: language pattern learning for word sense induction and disambiguation. Knowl.-Based Syst. **188**, 105017 (2020)
10. Harris, Z.S.: Distributional structure. WORD **10**(2–3), 146–162 (1954)
11. Hope, D.R.: Graph-based approaches to word sense induction. Ph.D. thesis, University of Sussex (2015)
12. Kuratov, Y., Arkhipov, M.: Adaptation of deep bidirectional multilingual transformers for Russian language (2019)
13. Lesk, M.: Automatic sense disambiguation using machine readable dictionaries: how to tell a pine cone from an ice cream cone. In: Proceedings of the 5th AICSD, pp. 24–26. ACM, New York (1986)
14. Lopukhin Konstantin, L.A.: Verbs using semantic vectors and dictionary entries. In: Computational Linguistics and Intellectual Technologies: Papers from the Annual conference "Dialogue", Moscow, Russia, pp. 393–405 (2016)
15. Luboshnikov, E., Makarov, I.: Federated learning in named entity recognition. In: van der Aalst, W.M.P., et al. (eds.) AIST 2020. CCIS, vol. 1357, pp. 90–101. Springer, Cham (2021). https://doi.org/10.1007/978-3-030-71214-3_8
16. Makarov, I., Bulanov, O., Gerasimova, O., Meshcheryakova, N., Karpov, I., Zhukov, L.E.: Scientific matchmaker: collaborator recommender system. In: van der Aalst, W.M.P., et al. (eds.) AIST 2017. LNCS, vol. 10716, pp. 404–410. Springer, Cham (2018). https://doi.org/10.1007/978-3-319-73013-4_37
17. Makarov, I., Gerasimova, O., Sulimov, P., Zhukov, L.E.: Dual network embedding for representing research interests in the link prediction problem on co-authorship networks. PeerJ Comput. Sci. **5**, e172 (2019)
18. Makarov, I., Kiselev, D., Nikitinsky, N., Subelj, L.: Survey on graph embeddings and their applications to machine learning problems on graphs. PeerJ Comput. Sci. **7**, e357 (2021)
19. Makarov, I., Makarov, M., Kiselev, D.: Fusion of text and graph information for machine learning problems on networks. PeerJ Comput. Sci. **7**, e526 (2021)
20. Mikolov, T., Chen, K., Corrado, G., Dean, J.: Efficient estimation of word representations in vector space. In: Proceedings of Workshop at ICLR 2013 (2013)
21. Miller, G.A., et al.: Introduction to wordnet: an on-line lexical database. Int. J. Lexicogr. **3**(4), 235–244 (1990)
22. Navigli, R., Ponzetto, S.P.: Babelnet: building a very large multilingual semantic network. In: Proceedings of the 48th Annual Meeting of the Association for Computational Linguistics, pp. 216–225 (2010)
23. Panchenko, A., et al.: RUSSE'2018: a shared task on word sense induction for the Russian language. In: "Dialogue", pp. 547–564. RSUH, Moscow (2018)
24. Pantel, P., Lin, D.: Discovering word senses from text. In: Proceedings of 8th ACM SIGKDD IC, pp. 613–619. ACM, New York (2002)

25. Peters, M.E., et al.: Deep contextualized word representations (2018)
26. Pugachev, A., Voronov, A., Makarov, I.: Prediction of news popularity via keywords extraction and trends tracking. In: van der Aalst, W.M.P., et al. (eds.) AIST 2020. CCIS, vol. 1357, pp. 37–51. Springer, Cham (2021). https://doi.org/10.1007/978-3-030-71214-3_4
27. Sowa, J.F., Siekmann, J.H.: Conceptual Structures: Current Practices. Springer-Verlag, Heidelberg (1994). https://doi.org/10.1007/3-540-58328-9
28. Struyanskiy, O., Arefyev, N.: Neural networks with attention for word sense induction. In: AIST (2018)
29. Tikhomirova, D., Nikitinskyb, N., Makarova, I.: Named entity recognition from chernobyl documentaries. In: Proceedings of EEML Workshop. Ceur WP (2020)
30. Tikhomirova, K., Makarov, I.: Community detection based on the nodes role in a network: the telegram platform case. In: van der Aalst, W.M.P., et al. (eds.) AIST 2020. LNCS, vol. 12602, pp. 294–302. Springer, Cham (2021). https://doi.org/10.1007/978-3-030-72610-2_22
31. Ustalov, D., Panchenko, A., Biemann, C., Ponzetto, S.P.: Watset: local-global graph clustering with applications in sense and frame induction. Comput. Linguist. **45**(3), 423–479 (2019)
32. Vaswani, A., et al.: Attention is all you need. ArXiv abs/1706.03762 (2017)

Computer Vision

Analysis of Thresholding Methods for the Segmentation of Brain Vessels

Alexey Kruzhalov$^{(\boxtimes)}$ and Andrey Philippovich

Moscow Polytechnic University, Moscow 107023, Russian Federation
alexkruzhalov@gmail.com
https://mospolytech.ru/en/

Abstract. In this paper, the problem of segmentation of the vascular network is solved based on the results of rotational angiography. This task is considered as one of the stages of data pre-processing in the construction of a computer-aided diagnostic system for the recognition of cerebral aneurysms. A statistical analysis of the effectiveness of using threshold segmentation methods to solve the problem under consideration was carried out and the choice of one of them was justified. The following methods are considered in this study: Otsu's method, Yen's thresholding method, triangle method, and Sauvola method. As a result of the experiments, it was determined that, among the considered methods, the triangle method, followed by the selection of the largest connected area, is the most suitable for solving the problem of vascular segmentation because this method proved to be the most sensitive to the isolation of low-contrast vessels of small size, allowing to get a complete picture of the vascular network structure.

Keywords: Medical image processing · Vessel Segmentation · Image thresholding

1 Introduction

An aneurysm is an abnormal bulge or bubble in the wall of a blood vessel. Studies show that cerebral vascular aneurysms occur in approximately 3–5% of healthy adults [4]. They often remain undiagnosed until clinical symptoms appear. Symptoms occur due to compression of the anatomical brain structures located next to the aneurysm, as well as due to the rupture of the aneurysm, leading to subarachnoid hemorrhage. Non-traumatic subarachnoid hemorrhage (NSAH) is one of the most severe and frequent forms of cerebral circulatory disorders. In the Russian Federation, the frequency of occurrence of NSAH is about 13:100,000 of the population per year [1]. The rupture of a cerebral aneurysm is associated with a high mortality rate (20–30% during the first 2–3 weeks after the rupture, 46% within 1 month [1]), and, therefore, a lot of research is currently being conducted to study the risk factors for aneurysm rupture, aneurysm development and growth. The risk of aneurysm rupture is affected by its location, size and shape [4].

E. Burnaev et al. (Eds.): AIST 2021, CCIS 1573, pp. 85–95, 2022.
https://doi.org/10.1007/978-3-031-15168-2_8

Timely diagnosis of aneurysms is possible with the methods of medical imaging: computed tomography and magnetic resonance imaging, rotational angiography. Early detection of the disease is based on the use of angiography. According to the data obtained, the doctor should not only detect the presence or absence of an aneurysm, but also assess its size and the risk of rupture. This is a rather time-consuming task, which can be partially automated due to the development and implementation in clinical practice of computer-aided diagnostics systems for recognizing aneurysms. When developing such a system, it is necessary to solve three tasks: to determine the presence of an aneurysm, to identify its boundaries (to perform segmentation) for the purpose of subsequent morphometric analysis, as well as to calculate the risk of rupture.

One of the stages of pre-processing of images of angiography results for subsequent recognition of cerebral aneurysms is the selection of an area corresponding to the vascular network. During angiography, a contrast agent is injected into the vessels, which allows them to be distinguished against the background of nearby tissues. The possibility to delineate blood vessels is provided by contrast enhancement of the image, which consists in increasing the intensity difference between the voxels corresponding to the blood vessels and the voxels corresponding to other tissues.

The task of segmentation of the vascular network is standard, and quite a lot of methods have already been proposed to solve it for various anatomical areas [11]. In particular, recent studies have proposed neural network algorithms for solving this problem [5,6,10], which have a sufficiently high accuracy. In earlier works, algorithms based on classical segmentation methods were proposed [2,13].

The complexity of building neural network algorithms lies in the fact that their training requires large-volume labeled data, and not all researchers have the opportunity to organize the process of data collection and markup. Classical segmentation algorithms do not require data for training, but many of them require interaction with the user (e.g., setting the starting points in the region growing method), which makes them inapplicable for automatic data processing.

In this paper, the segmentation of the vascular network is considered not as an independent task, but as one of the stages of data pre-processing for constructing a neural network algorithm for recognizing cerebral aneurysms. As a data preprocessing method, the vascular segmentation algorithm should meet the following criteria: it should not require user interaction (since data processing is required to be done automatically), and it should be as computationally simple as possible, as long as it produces acceptable segmentation results. Since we seek to reduce the overhead of processing each image while processing huge datasets, computational simplicity is crucial. Furthermore, the algorithm's computational complexity should be considered in the context of a hypothetical future deployment in clinical practice, as the algorithm's execution time will directly influence the efficiency of workflow automation. Furthermore, not all medical organizations have the resources to provide sophisticated computers to doctors' offices.

As the research results show, threshold segmentation methods may be utilized to successfully solve the problem of cerebral vessels segmentation [15]. Binarization as a preprocessing data approach can improve the accuracy of neural networks while tackling particular issues [7], and it is also frequently employed as the initial stage in image processing in more sophisticated data preparation techniques [9]. The speed of work and acceptable accuracy of the results acquired are two advantages of threshold segmentation methods in connection to the problem solution under consideration [15]. The lack of consideration for the properties of a selected region (for example, connectivity), in addition to the intensity values of the voxels, is one of the downsides, as is the difficulty in determining the threshold value for a specific image. The reason for the latter may be, for example, the inhomogeneous contrast of vessels in the image.

Based on the preceding considerations, we decided to investigate the efficacy of threshold segmentation methods concerning the problem of segmenting cerebral vessels based on rotational angiography results to select the best method for data pre-processing for use in developing an automated system for detecting cerebral aneurysms.

2 Related Work

The majority of recent studies devoted to tackling the challenge of cerebral vascular segmentation have used neural network techniques. Three-dimensional convolutional neural networks (3D CNNs), which are based on the U-Net architecture, are commonly applied. This design has shown to be effective in overcoming biomedical image recognition difficulties. The Dice coefficient (DSC) is the most commonly used metric for evaluating the precision with which a segmentation task is solved.

M. Livne et al. [10] created an artificial neural network model for recognizing brain arteries based on MRI scans, which had a $DSC = 0.88$. The model was trained on fragments of the original images. The authors discovered that the greater the context (i.e., the fragment) used in the studies, the higher the recognition accuracy. Based on the results of MR angiography, A. Hilbert et al. [6] introduced the BRAVE-NET neural network model for detecting arterial brain vessels, which obtained recognition accuracy of $DSC = 0.93$. One of the distinguishing aspects of the suggested technique is the simultaneous assessment of the spatial organization of data at many scales. The design of a neural network for detecting the head and neck vasculature based on CT findings was proposed by F. Fu et al. [5], and the validation set achieved $DSC = 0.94$.

Neural networks now prove beneficial in solving the problem at hand; however, as previously noted, working with them necessitates substantial computing resources and labeled data. Due to legal restrictions regarding protecting patients' personal data, the data utilized in the works mentioned above were not made publicly available. Furthermore, we could not identify a single dataset with a cerebrovascular network marker in the public domain that may be used to solve this challenge. This restricts a wide range of researchers from using these technologies. As a result, we can conclude that methods that do not require preliminary data markup are still relevant. Threshold segmentation methods, which are the subject of this research, are an example of such methods.

Threshold methods, despite their simplicity, can produce reasonably satisfactory results. The threshold algorithm described by R. Wang et al. [15] for segmentation of cerebral vessels, for example, has a recognition accuracy of $DSC = 0.84$. This shows that threshold approaches can be used as a preprocessing strategy for isolating the vasculature to detect cerebral aneurysms later on. The further processing of the received images can smooth out the lack of precision in the selection of the vascular network as compared to neural network techniques.

3   Methods

This paper uses the data published within the framework of the Cerebral Aneurysm Detection (CADA) Challenge [3]. It is devoted to the development of an algorithm for the recognition of cerebral aneurysms. The dataset includes 110 images. For each image, there is a segmentation mask of regions of interest made by an experienced neurosurgeon. During the reconstruction of the area of interest, 220 slices were generated with a matrix of 256×256 voxels. The voxel size was approximately 0.5 mm. The data were obtained with AXIOM Artis angiographic system of the C-arm type. The data were standartized before binarization.

Binarization methods based on adaptive threshold determination can be divided into global and local ones. Global methods determine the threshold value that is optimal in a certain sense for the entire image as a whole. Local methods consider different regions of the image separately, selecting an optimal threshold value for each of them. Local methods are significantly more computationally expensive compared to global ones. The global methods considered in this paper include the Otsu's method [12], Yen's thresholding method [16], the triangle method [17]. Among the local methods, the Sauvola method [14] is considered.

In this paper, the problem of segmentation of brain vessels is considered as one of the stages of data pre-processing for solving the problem of aneurysm recognition. We have segmentation masks for cerebral vascular aneurysms, so one of the criteria for evaluating the quality of solving the problem under consideration was the value of the intersection of the selected region with the region of interest (aneurysm). To switch from absolute values (the number of intersecting voxels) to a dimensionless metric, the volume of the intersection was calculated relative to the volume of the region of interest:

$$M_I = \frac{|V_T \cap V_{ROI}|}{|V_{ROI}|}, \tag{1}$$

where V_{ROI} is the set of voxels belonging to the region of interest, V_T is the set of voxels of the selected region.

The metric introduced above will not be sufficient for a comprehensive assessment of the quality of the problem solution, since in addition to maximizing the intersection of the selected region with the region of interest, we are interested in ensuring that the selected area covers the minimum possible number of voxels unrelated to vessels (i.e., there must not be false positives). We have no way to directly calculate the number of false positive voxels, since we do not have a segmentation mask of the vascular network. However, we can estimate the volume of the selected area compared to the area of interest. Let's introduce a metric M_O, which is the fraction of the selected area that does not belong to the region of interest:

$$M_O = \frac{|V_T/V_{ROI}|}{|V_T|}, \tag{2}$$

where V_{ROI} is the set of voxels belonging to the region of interest, V_T is the set of voxels of the selected region.

Since the area corresponding to the vascular network must be connected, it is proposed to use the selection of the largest connected area after binarization. In order to assess the statistical significance of the influence of the allocation of the largest connected region on the values of the target metrics, the Wilcoxon test was used. This is a nonparametric criterion used to assess the differences in the level of a certain quantitative indicator between two related (paired) samples. This case concerns dependent samples due to the fact that the analyzed binarization methods are applied to the same set of images.

In the Wilcoxon test, a set of values is calculated

$$D = \{ d_i = x_i - y_i \mid i = 1, ..., n \}, \tag{3}$$

where n is the number of objects in the sample, x_i is the value of the analyzed metric calculated for i-th object of the first sample, y_i is the value of the analyzed metric calculated for the i-th object of the second sample. The samples are sorted in such a way that the i-th object in the first and second samples is the same object (in this case, an image). The null hypothesis of the test is that the distribution of the analyzed metric in the considered samples is the same. In other words, the distribution of differences D is symmetric with respect to zero, i.e., the median of the resulting distribution is zero.

To analyze the results obtained for various thresholding methods, it is proposed to use the Friedman test. It is a generalization of the Wilcoxon criterion for the case when it is necessary to compare more than two related samples. The null hypothesis of the test is that there are only random differences between the measurements obtained under different conditions. An alternative hypothesis is that there are statistically significant differences between at least one pair of samples among the considered samples. To determine which samples have statistically significant differences between them, it is necessary to conduct pairwise comparisons using the Wilcoxon criterion. At the same time, it is necessary to take into account the effect of multiple comparisons, according to which, when checking a certain family of hypotheses, the probability of a type I error (false rejection of the null hypothesis) increases with an increase in the number of hypotheses being tested according to the formula $1 - (1 - \alpha)^m$, where m is the number of hypotheses being tested, and α is the selected significance level. In this paper, the classical correction for multiple comparisons by the Bonferroni method is used to preserve the probability of a type I error at level α.

4 Results and Discussion

The studies were conducted on a computer with an Intel Core i3-6100 CPU 3.70 GHz × 4 and 16 GB of RAM. Python 3.7 and Jupyter Notebook were used to implement the proposed computational experiments. We use the following additional software libraries: TorchIO (for loading and preprocessing data), SciPy (statistical tests), NumPy, scikit-image (binarization methods), and Matplotlib. 3D Slicer [8] was used to visualize the results obtained.

Table 1 shows the results of applying various binarization methods to the analyzed data. The column "Wilcoxon Test" contains the values of test statistic (W) and their corresponding p-values calculated for each method according to the metric M_I. The samples to be compared are the samples obtained with and without the use of the selection of the largest connected region.

Table 1. The results of the application of various binarization methods.

Method	Selection of the largest connected region	$\overline{M_I}$	$\overline{M_O}$	Wilcoxon test
Otsu's method	Yes	0.72	0.94	$W = 80,$
	No	0.73	0.95	P-value = 0.35
Yen's method	Yes	0.53	0.92	$W = 46,$
	No	0.57	0.93	P-value = 0.009
Triangle method	Yes	0.93	0.98	$W = 58$
	No	0.93	0.99	P-value = 0.08
Sauvola method	Yes	0.85	0.999	$W = 24$
	No	0.86	0.999	P-value = 0.02

The analysis of the results obtained shows that satisfactory results for the intersection metric ($\overline{M_I} > 0.8$) were obtained only when using the triangle method and the Sauvola method.

To analyze the statistical significance of the influence of the selection of the largest connected region, we set the significance level $\alpha = 0.05$. Analyzing the p-values obtained for the M_I metric, we can conclude that for the results obtained using the Otsu's method and the triangle method, we have insufficient grounds for rejecting the null hypothesis. This means that for these methods, the selection of the largest connected region does not significantly affect the value of the intersection of the selected region with the region of interest. For the Yen's method and the Sauvola's method, on the contrary, we have sufficient grounds for rejecting the null hypothesis, which means that for these methods, the determination of the maximal connected region has a significant impact on the value of the target metric, and this influence is negative, i.e., the target metric decreases. Statistically significant results were obtained for the metric M_O in all experiments: the selection of a connected region contributes to a statistically significant decrease in the M_O value.

For further comparison of thresholding methods in order to select the best, we will use the values of the target metrics obtained using the selection of the largest connected region. For a visual interpretation of the results obtained, we will depict the metric distributions calculated for different methods using a box plot (Fig. 1).

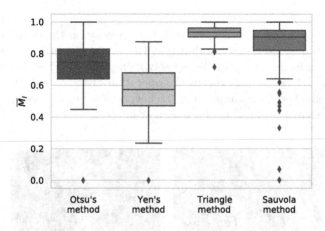

Fig. 1. Box plot of the intersection metric.

The diagram shows that the largest median value and the smallest spread of values relative to it are obtained for the triangle method, and the largest number of outliers is obtained for the Sauvola method.

To assess the statistical significance of the differences between the metrics M_I obtained using 4 different methods, the Friedman test was used. For the analyzed data, the following test statistic and p-value were obtained: $S = 223.32$, p-value $= 3.85e{-}48$. At the level of significance of $\alpha = 0.05$, there are sufficient grounds for rejecting the null hypothesis. In order to find out which methods produce statistically significant differences between their results, pairwise comparisons were performed (Table 2).

Based on the results obtained, we can say that statistically significant differences are present between all pairs of the considered methods. Since the triangle method showed the best results, and according to the results of pairwise comparisons, the differences between this method and all the others are statistically significant, we can conclude that the use of this method for the solution of the problem is the most preferable.

Table 2. Results of pairwise comparisons.

	Otsu's method	Yen's method	Triangle method
Yen's method	$W = 173$ P-value = 1.7e−16 P-value (adj) = 1.02e−15		
Triangle method	$W = 57$ P-value = 9.1e−19 P-value (adj) = 5.4e−18	$W = 0$ P-value = 1.3e−19 P-value (adj) = 7.7e−19	
Sauvola method	$W = 999$ P-value = 1.5e−09 P-value (adj) = 9.1e−09	$W = 384$ P-value = 2.8e−15 P-value (adj) = 1.7e−14	$W = 777$ P-value = 1.1e−07 P-value (adj) = 6.8e−07

In the analysis given earlier, we relied exclusively on the values of metrics. However, it should be borne in mind that the values of metrics are not always an objective indicator of the quality of solving a problem. Therefore, to confirm the results obtained, an analysis of the operation of the above algorithms was carried out on a specific example (Fig. 2). For the analysis, a slice of the original image was taken, which included a region of interest.

Fig. 2. Test image (on the left: the original image, on the right: the mask).

Figure 3 shows examples of the operation of the analyzed algorithms on a test image. The first line shows the results before the selection of the largest connected area, and the second line shows the results after that.

From the examples given, it can be seen that the Otsu's and Yen's methods identified only large vessels with high contrast. The triangle method turned out to be more sensitive to the selection of less contrasting small vessels, but when using it, some noise components are also highlighted, which are removed after the selection of the largest connected area. The Sauvola method showed a large number of false positives, so it was decided to exclude it from further consideration.

The images shown in Fig. 3 allow us to evaluate the work of the algorithms on the example of a single slice; however, they do not give an idea of the overall

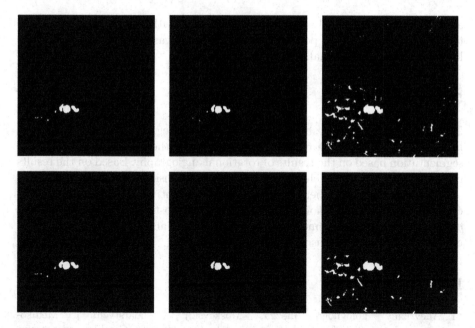

Fig. 3. The result of applying various binarization methods to the analyzed image (from left to right: the Otsu's method, the Yen's method, the triangle method; from top to bottom: without selection of the largest connected area and with its selection).

structure of the selected area. To evaluate the structure of the selected region in the volume, 3D reconstructions of the regions obtained based on various algorithms were constructed using the 3D Slicer program [8] (Fig. 4). In general, we can say that they confirm the previously made observations that the Otsu's

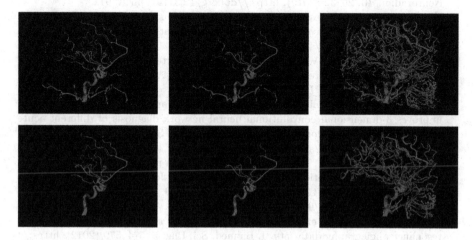

Fig. 4. 3D reconstructions of regions obtained with different binarization methods (from left to right: the Otsu's method, the Yen's method, the triangle method; from top to bottom: without selection of the largest connected area and with its selection).

and Yen's methods mainly delineate large vessels with high contrast, and the triangle method makes it possible to select a wider vascular network, including low-contrast vessels of small size.

5   Conclusion

The article analyzed threshold methods of image segmentation (binarization) in order to determine the method that best solves the problem of vasculature segmentation based on the results of rotational angiography. Based on the results of the analysis, it was concluded that, among the considered methods, the most suitable is the triangle method with the subsequent allocation of the largest connected area. This method turned out to be the most sensitive to the isolation of low-contrast vessels of small size, and, therefore, it allows for obtaining a more complete picture of the vascular network structure.

References

1. Anevrizmy golovnogo mozga, Moskovskij Gosudarstvenny'j Mediko-Stomatologicheskij Universitet imeni A.I. Evdokimova (Cereblar aneurysms, Moscow State University of Medicine and Dentistry). www.myuniverclinic.ru/articles/anevrizmy-golovnogo-mozga/. Accessed 10 May 2021
2. Babin, D., Piourica, A., Vylder, J.D., Vansteenkiste, E., Philips, W.: Brain blood vessel segmentation using line-shaped profiles. Phys. Med. Biol. **58**, 8041 (2013). https://doi.org/10.1088/0031-9155/58/22/8041
3. CADA - Cerebral Aneurysm Detection. https://cada.grand-challenge.org/. Accessed 9 Aug 2021
4. Faron, A., Sijben, R., Teichert, N., Freiherr, J., Wiesmann, M., Sichtermann, T.: Deep learning-based detection of intracranial aneurysms in 3D TOF-MRA. Am. J. Neuroradiol. **40**, 25–32 (2019). https://doi.org/10.3174/ajnr.A5911
5. Fu, F., et al.: Rapid vessel segmentation and reconstruction of head and neck angiograms using 3D convolutional neural network. Nat. Commun. **11**, 1–12 (2020). https://doi.org/10.1038/s41467-020-18606-2
6. Hilbert, A., et al.: BRAVE-NET: fully automated arterial brain vessel segmentation in patients with cerebrovascular disease. Front. Artif. Intell. **3**, 1–14 (2020). https://doi.org/10.3389/FRAI.2020.552258
7. Ker, J., Singh, S.P., Bai, Y., Rao, J., Lim, T., Wang, L.: Image thresholding improves 3-dimensional convolutional neural network diagnosis of different acute brain hemorrhages on computed tomography scans. Sensors **19**(9), 2167 (2019). https://doi.org/10.3390/s19092167
8. Kikinis, R., Pieper, S.D., Vosburgh, K.G.: 3D slicer: a platform for subject-specific image analysis, visualization, and clinical support. In: Jolesz, F.A. (ed.) Intraoperative Imaging and Image-Guided Therapy, pp. 277–289. Springer, New York (2014). https://doi.org/10.1007/978-1-4614-7657-3_19
9. Kim, D.-Y.: 3D volume extraction of cerebrovascular structure on brain magnetic resonance angiography data sets. J. Biomed. Sci. Eng. **5**, 574–579 (2012). https://doi.org/10.4236/jbise.2012.510070

10. Livne, M., et al.: A U-net deep learning framework for high performance vessel segmentation in patients with cerebrovascular disease. Front. Neurosci. **13**, 97 (2019). https://doi.org/10.3389/fnins.2019.00097
11. Moccia, S., De Momi, E., Hadji, S.E., Mattos, L.S.: Blood vessel segmentation algorithms-Review of methods, datasets and evaluation metrics. Comput. Meth. Program. Biomed. **158**, 71–91 (2018). https://doi.org/10.1016/j.cmpb.2018.02.001
12. Otsu, N.: A threshold selection method from gray-level histograms. IEEE Trans. Syst. Man Cybern. **9**(1), 62–66 (1979). https://doi.org/10.1109/TSMC.1979.4310076
13. Passat, N., Ronse, C., Baruthio, J., Armspach, J.-P., Maillot, C., Jahn, C.: Region-growing segmentation of brain vessels: an Atlas-based automatic approach. J. Magn. Reson. Imaging **21**, 715–725 (2005). https://doi.org/10.1002/JMRI.20307
14. Sauvola, J., Pietikäinen, M.: Adaptive document image binarization. Pattern Recognit. **33**, 225–236 (2000). https://doi.org/10.1016/S0031-3203(99)00055-2
15. Wang, R., et al.: Threshold segmentation algorithm for automatic extraction of cerebral vessels from brain magnetic resonance angiography images. J. Neurosci. Meth. **241**, 30–36 (2015). https://doi.org/10.1016/J.JNEUMETH.2014.12.003
16. Yen, J.C., Chang, F.J., Chang, S.: A new criterion for automatic multilevel thresholding. IEEE Trans. Image Process. **4**, 370–378 (1995). https://doi.org/10.1109/83.366472
17. Zack, G.W., Rogers, W.E., Latt, S.A.: Automatic measurement of sister chromatid exchange frequency. J. Histochem. Cytochem. **25**, 741–753 (1977). https://doi.org/10.1177/25.7.70454

Transformer-Based Deep Reinforcement Learning in VizDoom

Vitalii Sopov[1] and Ilya Makarov[1,2(✉)] [iD]

[1] HSE University, Moscow, Russia
`iamakarov@hse.ru`
[2] Artificial Intelligence Research Institute (AIRI), Moscow, Russia

Abstract. Transformers is a novel neural network architecture that is successfully used in natural language processing tasks and is starting to be used in other areas such as video processing and image processing. However, transformers are yet to be studied in different aspects of reinforcement learning scenarios. In this work we combine transformer architectures with reinforcement learning and train them in the VizDoom game environment, producing agents that play better in comparison to traditional neural network architectures.

Keywords: Deep Reinforcement Learning · Transformer · Convolutional neural networks · VizDoom

1 Introduction

Imitating human behavior is challenging tasks related to Turing test for AI in Games [27, 28, 30–32]. Considering first-person 3D shooters, it is even more difficult to build believable agent achieving human-level performance: reactions to the correct behavior are usually extremely rare and are very delayed. This means that the agent can receive a reward signal for an action that he performed hundreds of frames back. The information received by the agent from the environment is limited by a first-person view. This is opposite to the settings of team games [29] or casual games [17].

Reinforcement learning (RL) is an area of machine learning which studies how intelligent agents operate in different environments to maximize their cumulative reward. The reward is a scalar which usually correlates with the intuitively successful agent's actions such as surviving as long and killing as many enemies as possible in shooter games or beat-them-ups, or driving through the track faster in racing games, or running farther in runners, or even using the robot manipulator in a real-world as fast and precisely as possible, etc.

Many modern RL algorithms must store the data concerning game states and their possible transitions between each other. However, in most possible tabletop and computer games, this is impossible: for example, in a game of "Go" there are about $2.08 \cdot 10^{170}$ legal positions, which is unfeasible for modern computers.

ⓒ The Author(s), under exclusive license to Springer Nature Switzerland AG 2022
E. Burnaev et al. (Eds.): AIST 2021, CCIS 1573, pp. 96–110, 2022.
https://doi.org/10.1007/978-3-031-15168-2_9

When emulating human players in computer games, RL algorithms have access only to the images of the game screen, and even for small 64×64 images their whole set cannot be stored and processed. The modern solution for this problem is to process and approximate game states using deep neural networks, thus creating the so-called Deep Reinforcement Learning.

Nowadays most neural networks used in reinforcement learning are based on traditional architectures, such as convolutional neural networks or recurrent neural networks. However, recently a new novel neural network architecture called transformers emerged. It became widely used in natural language processing and other tasks involving the processing of sequences, outperforming recurrent networks, the architecture which had previously dominated in sequence processing tasks. Since in reinforcement learning we can work with short sequences of most recent game frames, this is a promising domain to use transformers in. In this work we combine parts of transformer architecture with traditional neural network models for reinforcement learning, trying to improve the usage of the A2C algorithm [48] via the special properties and strengths of transformers while keeping the opportunity to generalize to other deep RL algorithms.

The contributions of this work are a novel transformer-based architecture for reinforcement learning in VizDoom game environment; the review of existing modifications useful for stabilization of transformers in reinforcement learning; proposed novel modifications comparison in several well-known VizDoom scenarios, showing improvements in performance.

The structure of the work is as follows. Firstly we describe the formal problem statement in reinforcement learning, then make a literature review of related works both in reinforcement learning and transformer models. Then, we propose our transformer-based model architecture and its modifications comparing them with traditional convolutional and recurrent architectures on well-known VizDoom scenarios and concluding the results.

2   Preliminaries

In this section, we make a brief overview of the formalization of the task we solve. Markov decision process (MDP) is a mathematical model which formalizes the decision-making process of an agent and its consequences in environments in which the outcomes are not fully controlled by the agent. At each time step, the process is in some state s, and the agent is free to choose any action a from the list of actions available in this state. The process is then moved to another state s', and the agent receives reward r. More formally, Markov decision process is defined by a tuple (S, A, P_a, R_a), where S is a set of all states, A is a set of all possible actions, $P_a(s, s') := Pr(s_{t+1} = s' | s_t = s, a_t = a)$ is the probability of the next state being s_t when taking action a at state s, $R_a(s, s')$ is the reward received by transitioning from s to s' by taking action a. Agent's policy is usually defined as a stochastic function π which returns the probability distribution of the next action at a given state.

The quality of action performed at the specific state depends on the cumulative reward which agent gets by continuing the process from the given state-action pair. Its value is named Q (also called Q-value or action value) and is formalized by the following equation:

$$Q(s,a) = R_a(s,s') + \mathbb{E}[\sum_{t=1}^{+\infty} \gamma^t R_{a_t}(s_t, s_{t+1})],$$

where a_t, s_t are mean actions and states at t-th step after the initial transition, and $\gamma \in [0; 1]$ is the discount factor meaning that events in the far future are usually less important to the agent than nearer ones.

This equation can be reformulated in a recursive way which is called the Bellman equation:

$$Q(s,a) = R_a(s,s') + \sum_{a'} \gamma \cdot \pi(a'|s') \cdot Q(s',a').$$

State value defines the quality of the given state, and its function is named V (or V-value) and is defined as

$$V(s) = \mathbb{E}[\sum_{t=1}^{+\infty} \gamma^t R_{a_t}(s_t, s_{t+1})|s_0 = s]$$

The Bellman equation for V is defined in this way:

$$V(s) = \sum_a \pi(a|s) \cdot \mathbb{E}[R_a(s,s') + \gamma V(s')]$$

Many popular reinforcement learning algorithms (mainly Q-learning [46], Deep Q-learning [36] and its derivatives [35,48] operate in terms of Q- and V-functions and estimate them to construct their policies.

Markov Decision Process implies that full information about the current state is available. In many environments though, such as for example most 3D games, the agent does not have access to all the information about the state, only having observations (such as a game screen). Partially observable Markov decision process (POMDP) introduces Ω, which is a set observations, and $O_a(s',o) = P(o_{t+1} = o|s_{t+1} = s', a_t = a)$ is the probability of the next observation being o' when taking action a and moving to state s'. The agent operates over this mapping, and a POMDP can be formulated as an MDP with observations viewed as process states [16].

3   Related Work

In this section, we make an overview of the key reinforcement learning publications, list works related to transformers architecture, and describe several articles that combine reinforcement learning and transformers in other environments.

Originally the first work to combine modern convolutional neural networks and reinforcement learning (particularly Q-learning) was written by Mnih et al. and proposed the approach called deep Q-learning [36]. It used neural networks to approximate states in high-dimensional spaces and allowed the usage of reinforcement learning in large state sets which was previously impossible due to limitations in storage and computational power. Many more recent papers are based on this work, directly or indirectly.

A popular approach in deep reinforcement learning is to decouple estimation of the state's value and computation of the action which the agent should take. It created a family of actor-critic algorithms which train separate neural networks or a single network with multiple heads to implement this idea [10]. Another paper by Mnih et al. [35] proposed an approach called Asynchronous Advantage Actor Critic, or A3C. It took previously known Advantage Actor Critic and parallelized it, running multiple agents in parallel and updating the model's weights asynchronously by each agent.

Synchronous Advantage Actor Critic [48] is a modification of A3C that gathered losses over all agents and updated the model's weights synchronously, which was both more simple than the original A3C and also resulted in more stable training and faster convergence, also improved in several applications of deep RL [3,26]. We use A2C as the underlying algorithm of choice because, while more modern and sometimes more efficient algorithms exist, it is simple, has relatively good performance, and is well-known in comparison to more recent works.

3.1 VizDoom

This work uses VizDoom as the RL environment of choice. VizDoom [20] is an open-source implementation of 3D shooter Doom which is focused on allowing researchers to programmatically control the game and the agent and simultaneously receive both visual information such as game screen and also internal metrics such as player statistics, coordinates of player and enemies, depth buffer and other parameters. The novelty of this game in comparison to other environments is that the agent operates in 3D space, while most other popular environments are platformers operating in 2D space, such as most games on Atari [5].

The game provides a set of scenarios that define different tasks for agents to solve: crossing the corridor full of enemies, finding the way through the maze, collecting health kits, defending the point against waves of enemies, or battling on the arena full of different enemies, weapons and boosts. Custom scenarios can also be made. This allows researchers to test different scenarios and train different skills, such as 3D mapping and navigation [1,13,51], exploration of new behaviors [38], curriculum learning [49], or generalization to other yet unseen problems [47].

3.2 Transformers

Transformers [44] is a novel neural network architecture that utilizes the mechanism of attention to apply different degrees of importance to different parts of input data. Such networks are designed to handle sequential data and use various ways of positional encoding to capture the context of the individual parts of the input. This architecture was primarily used to handle textual data and corresponding tasks such as text translation, summarization, named entity recognition and others. One of the most known applications of this architecture is the BERT model [11], which is a transformer trained on huge amounts of text data. This specifically trained model gives state-of-the-art performance in neural translation and other natural language processing tasks, greatly outperforming other models, although for the price of its large size and huge computational costs both for training and inference.

Later it was adopted to other fields such as processing of video sequences for segmentation, summarization, and other video-related tasks [6,45]. There were successful attempts to apply transformers to non-sequential input data such as images, such as a Vision Transformer model [12]. The images, in that case, were split into square patches which were then considered sequence tokens and processed as in regular transformer architectures, and the model showed considerable improvements on an ImageNet dataset in comparison to traditional convolutional networks.

Many recent works make comparisons of transformer-based models with traditional architectures in various tasks, usually with transformers outperforming convolutional [39,42] and recurrent [8,22,34] models in image and sequence processing, respectively. To the best of our knowledge, there have not been any attempts to use transformers in deep reinforcement learning tasks in VizDoom environment, despite certain attempts to apply transformers in deep RL settings for text-based games, simple environments or modelling state-action sequences [7,43,50]. Among the notable ones, a multi-agent reinforcement learning model proposed by S. Hu et al. [14] used self-attention mechanism to tune action importance of agents' actions when solving various tasks in StarCraft Multi-Agent Challenge [40]. Another single-agent reinforcement learning model by E. Parisotto et al. [37] proposed several stabilizing modifications of transformers and successfully compared them with traditional recurrent neural networks in various tasks from DMLab-30 benchmark suite [4].

4   Model Architecture

In this section, we describe the structure of our transformer-based model and its traditional alternatives with which we will later compare our model, list existing modifications to the original architecture proposed in other publications and describe our own techniques.

Our agent is based on the previously mentioned A2C model [48] and on its implementation by Pratyush Kar [19]. The model consists of several consecutive convolutional layers, a transformer block consisting of 2 layers, a common dense layer, and then separate dense heads for actor and critic submodels required for the A2C algorithm.

The transformer block implements the original transformer architecture only for the encoder part since we do not need a sequence decoder. The input to the transformer is provided as a series of N last consecutive frames concatenated vertically into a single picture. It is then split into N separate frames, which are passed through the convolutional layers and then flattened and combined into a single tensor. This tensor is presented by a special learnable "CLS" token. The resulting tensor is then passed to all encoder layers consecutively, and the transformed output of the last token in the sequence corresponding to the most recent frame is then fed into the next model's layers.

We compare our transformer model with two other architectures: a classical convolutional network and a recurrent neural network utilizing LSTM layers [9]. Both these models contain the same dense heads for Actor and Critic submodels. In the LSTM model, transformer blocks are replaced with LSTM layers, the convolutional model consists of a larger number of differently shaped convolutional layers. The exact configuration and parameters of layers were tuned manually, both for better performance and to maintain the comparable model size. The exact configurations of layers and their parameters are described in Table 1.

Table 1. Transformer (left), LSTM (middle), and Convolutional (right) encoders.

Layer	Parameters
Conv2D	3@7x7, stride = 4
ReLU	–
Conv2D	8@5x5, stride = 3
ReLU	–
Flattening + PosEncoding	–
TransformerLayer	attention_heads = 2, output_size = 288
TransformerLayer	attention_heads = 2, output_size = 288
Dense	512
ReLU	–
Dense, 1	Dense, ACTIONS_NUM Softmax

Layer	Parameters
Conv2D	3@7x7, stride = 4
ReLU	–
Conv2D	8@5x5, stride = 3
ReLU	–
LSTMLayer	output_size = 288
LSTMLayer	output_size = 288
Dense	512
ReLU	–
Dense, 1	Dense, ACTIONS_NUM Softmax

Layer	Parameters
Conv2D	32@8x8, stride = 4
ReLU	–
Conv2D	64@4x4, stride = 2
ReLU	–
Conv2D	32@3x3, stride = 1
ReLU	–
Flattening	
Dense	512
ReLU	–
Dense	512
ReLU	–
Dense, 1	Dense, ACTIONS_NUM Softmax

4.1 Applicability of Transformers to Deep RL

In addition to the original transformer architecture, several modifications were applied to the model and compared between each other. Most of these modifications were taken from other works and propose either an adaptation of transformers for working with image data or a way to stabilize transformers for the domain of reinforcement learning.

The first modification "Reordered LayerNorm" (see Fig. 1) was the reordering of LayerNorm blocks in the encoder layer. This modification was proposed [37], and the main motivation is that it allows an identity map from the input of the first encoder layer to the output of the last layer. Coupled with the fact that layers are initialized with near-zero values, encoder layers pass the input reservations to the Actor and Critic heads more or less unchanged, allowing them to learn Markovian policy at the beginning, learning some simple reactive behaviors much more quickly. Also, the reordering of LayerNorm blocks is also used in Vision Transformers, [12] and thus when training ViT-based models, we also imply the usage of this modification.

The second modification "Gated Transformer" (see Fig. 1) was the usage of so-called Gated Transformers [37] which reportedly should have stabilized the model and improved its total performance by the usage of the gating technique from GRUs (Gated Recurrent Units). This gating mechanism replaces original residual connections and reportedly increases the stability of the model.

The third modifications "ViT, 4 20 × 20 patches" and "ViT, 16 10 × 10 patches" (see Fig. 1) was to replace traditional transformers with Vision Transformer proposed in [12]. In the base model, the input images are fed into a sequence of convolutional layers and then flattened, producing a single flat vector corresponding to the whole frame. This encoded frame would then be considered a single transformer token, thus the whole input sequence for the transformer would be defined using the temporal dimension. Vision Transformers split the input images into a square grid of image patches (for example, for an image that is 256 × 256 px it would produce 64 patches which are 32 px both in height and width). A single patch then would be considered a separate token, thus providing the spatial dimension. Since we consider not a single frame, but a sequence of last N frames, we should use both spatial and temporal information when applying positional encoding. In this work, we just concatenate the sequences of patches from consecutive frames, although separate positional embeddings could possibly be computed and combined (via a simple element-wise summation or other means).

The last simple modification "Without CLS-token, LayerNorm" (see Fig. 1) is proposed by us and is focused on removing the CLS token which is prepended to the input sequence that is fed into the transformer. Prepending this token is the originally intended way to solve classification tasks using transformers. The encoded representation of this token is fed into a feed-forward head or heads which are then trained for classification or regression tasks. This approach is used in BERT [11], mentioned in Vision Transformers article [12] and used in at least one of its implementations [15]. However, this token increases the total sequence length and thus makes the model size larger, and, in our case, its usage is arguably not necessary. While for text classification tasks applied to sequences of words or characters or image classification tasks in Vision Transformers we should not focus on individual words or image patches, thus introducing a separate CLS token that is detached from all other tokens in the input sequence can be useful. However, in our task, the last element of the sequence (the most recent frame) most likely contains the most useful information for the agent.

4.2   Ablation Study on Transformer-Based Modifications

For comparison of modifications, only a select few scenarios were used, mainly "Health Gathering" and "My Way Home". This choice was made mainly because of computational limitations. However, in most cases, it was clear if a modification has brought any improvements or not. The results are shown in Fig. 1.

The reordering of LayerNormalization blocks in encoder proposed in [37] showed significant improvements over the unmodified transformer model, with a huge gap in performance. As this modification is used in Visual Transformers

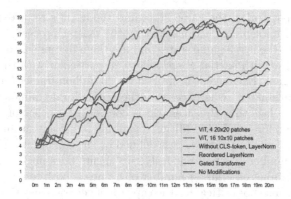

Fig. 1. Mean reward for modifications, health gathering

and is also coupled with the Gated Transformer modification in its paper [37], we consider this the default modification and combine all other modifications with it when it is not stated otherwise. This mainly concerns the CLS-token modification, since other modifications already use LayerNormalization reordering.

Vision Transformers were discovered to be rather tricky to train, and their usage is heavily limited by GPU memory constraints. Smaller patch sizes (such as 4×4, 5×5, etc.) result in worse training efficiency significantly. Larger patch sizes improve the performance, but at the same time increasing patch sizes and decreasing the number of patches defeats the idea of Vision Transformers and extrapolates to using one single patch size, thus falling back to the original transformer architecture. Besides, although transformers with larger sizes trained far better, the training itself was slower since each image patch was large in size.

Gated Transformers did not bring the supposed stabilizing effect as was described in [37]. Although the model was trainable and showed improvement over its freshly initialized state, the progress was significantly slower than for other modifications, the training was interrupted at 20 million game steps.

The removal of CLS-token proposed in this work actually brought improvements, although they were not huge. Even if the model did not show any significant leaps in the resulting agent's performance, it sped up the training process and decreased the model size (by 25%, if we train on 3-frame tuples), freeing GPU memory which would allow us to increase the number of workers, for example, if it were our intention.

5 Training and Experiments

In this section, we describe the training process, its most important parameters, and the scenarios in which we will train our model.

In this work, we use a mostly unmodified A2C algorithm as described in [48] and implemented in [19]. For training we start 240 simultaneous worker processes, each of them runs a single independent VizDoom environment, sends the game data (in particular, the game state and the agent's rewards) to the master process, and implements actions sent by the master process. For each game episode, we store the last 3 game states and use them in tandem to capture game dynamics and deal with partial observability, in particular with the limited field of view of the agent. Each agent is trained for 20 million frames for all scenarios except the "Deathmatch" where we train the agents for 40 million frames. That is because of the fact that this scenario is considered much more difficult and complex and therefore requires additional training to get any meaningful results. Both this and the number of workers is defined by the computational limitations, such as larger number of simultaneously run agents does not fit into the GPU memory, and training on a larger number of frames is both too computationally expensive and is also not critical for the capture of general trends in experiments.

The experiments were conducted on several default VizDoom scenarios. In "Health Gathering", the agent is placed in a rectangular room filled with randomly spawning health kits. The room's floor is toxic and decreases the player's health every second, thus the agent is motivated to collect health kits to survive as long as possible, for which it is rewarded. In "My Way Home", the agent is placed in a random room of a simple maze consisting of several differently looking rooms. The goal is to find an armor item that gives the agent a huge reward and ends the episode. The agent is motivated to find the armor as fast as possible because it gets a small penalty every second until it finds the armor. In "Defend the Center", the agent is spawned in the center of a circular room and can only fire its pistol, turn left and right. Monsters spawn at random locations in the room, and the goal of the agent is to kill as many foes as possible until it is either killed or spends all its ammunition. In "Deadly Corridor", the agent is required to pass a long corridor and reach the armor item. The task is complicated by monsters spending at the sides of the corridor, and the agent must kill them first or it will die itself. "Deathmatch" is the most complex scenario and it incorporates many elements from other scenarios. Its map consists of a large square arena and four rooms on each side of it. The arena is mostly empty, and each room contains various weapons, ammunition, and useful items. The agent appears at a random point of the arena, after that in random time intervals monsters of various types spawn at random locations. The goal of the agent is to kill as many monsters as possible using the items it collects.

6   Results

In this section, we select the best set of modifications and then describe its performance against traditional architectures in several well-known VizDoom scenarios, both quantitatively and qualitatively.

For each scenario, we draw two plots: one with mean reward and one with maximal reward. Points on both plots indicate means and maximums over previous 76800 game frames over all 240 workers. Exponential smoothing with a smoothing coefficient 0.8 was applied. On each plot only three models are displayed: a convolutional model, a recurrent model, and a transformer-based model. Among all proposed modifications, the removal of CLS-token was used. Figure 2 presents results for "Health Gathering", "My Way Home", "Defend the Center", "Deadly Corridor" and "Deathmatch" scenarios, respectively.

From the results it can be seen that the transformer-based model significantly outperforms the convolutional network in all the scenarios (Fig. 2(a, b, e, f, g, h)) except for "My Way Home" and is usually better than the LSTM-based model, loosing only in "My Way Home" and "Deathmatch" scenarios.

In "Health Gathering" (Fig. 2(a, b)), the agents show similar behaviors, although the transformer model behaves less clumsily than others, while the convolutional model sometimes shows signs of non-optimal behavior, missing nearest health kits, even when they are on the way to other, larger clusters. LSTM-based model is less susceptible to this behavior. In "Defend the Center" (Fig. 2(e, f)) and "Deadly Corridor" (Fig. 2(g, h)), transformer-based agent expresses more economy in ammunition management than other models, while other behavior patterns seem similar.

In the case of "My Way Home" (Fig. 2(c, d)), the agent does not improve its performance at all during the first 20 million game steps (it slowly starts improving after that, but this is now shown on charts, and even at 30 million steps its performance is significantly worse than that of other models). What is interesting, only for this scenario our convolutional model outperforms both recurrent and transformer networks. Qualitatively, the convolutional model almost always successfully reaches the finish, although it does it rather clumsily, sometimes ignoring the goal when it is in the agent's direct view, however, the agent always reaches the goal in the end. The LSTM-based model mostly behaves less clumsily than the convolutional model and usually reaches the finish more quickly, but it gets stuck rather often, not finishing the level at all. Most likely additional training can fix this problem. On the other hand, a transformer-based agent behaves erratically, rarely showing signs of useful behavior, although seeing the goal in the direct view of the agent almost always leads to success. The hypothesis is that this is connected with the randomness of this scenario: the agent is randomly spawned at a random point inside of a maze, and its optimal strategy highly depends on this point, thus making the environment too unstable. This peculiarity significantly hinders the LSTM-based model and prevents the transformer-based model from being trained at all. However, this hypothesis should be checked on a series of additional experiments on specifically created scenarios in further works

In "Deathmatch" (Fig. 2(i, j)), the transformer-based model shows a higher speed of training at the beginning, but then quickly loses to the LSTM-based model, when the latter quickly leaps in terms of performance, drastically outperforming the transformer model. At the same time, the transformer still significantly outperforms the convolutional model. In terms of qualitative behaviors, all agents do not show policies that are close to optimal ones, although recurrent and transformer models try searching for weapons and power-ups and using them in combat, although they do not always do it successfully or smartly. It is

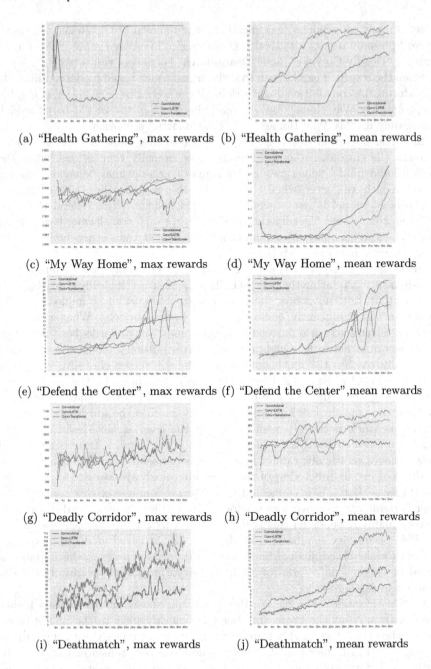

(a) "Health Gathering", max rewards (b) "Health Gathering", mean rewards

(c) "My Way Home", max rewards (d) "My Way Home", mean rewards

(e) "Defend the Center", max rewards (f) "Defend the Center", mean rewards

(g) "Deadly Corridor", max rewards (h) "Deadly Corridor", mean rewards

(i) "Deathmatch", max rewards (j) "Deathmatch", mean rewards

Fig. 2. Comparison of models in different scenarios

assumed that the agents must have additional training at a larger scale if even 40 million episodes are not enough for good results.

7 Conclusion and Further Work

Transformers turned out to be capable of providing improvements for reinforcement learning agents, in most cases outperforming traditional convolutional and recurrent architectures. However, not in all scenarios, the transformer-based agent was capable of outperforming other models or training at all, suffering in unstable and heavily randomized environments, and not all proposed modifications were helpful in improving performance. Thus, other stabilization techniques and modifications of transformers should be considered and applied. Other, less straightforward training techniques can be explored, such as unsupervised or semi-supervised pretraining, which are widely used when training transformers for image processing or natural language tasks [11,12].

More thorough experiments on the same scenarios and other tests on other scenarios can be conducted since in some scenarios (as, for example, "Deathmatch") agents can be trained for days to converge to best-possible policies, which was unfeasible in our work.

Besides, other deep RL algorithms should be used in conjunction with transformer architectures. In particular, most state-of-the-art models in reinforcement learning nowadays utilize distributed learning, such as MuZero, R2D2, and Agent57 agents from DeepMind [2,18,41]. This can be coupled with the supposed ability of transformer architectures to easily scale to huge numbers of parameters (similar to BERT trained on clusters of GPUs [11]), and usage of distributed learning can greatly contribute to the improvements in the performance of transformer-based architectures. In addition, our previous experiments on depth reconstruction [21,23–25,33] may substitute available information on depth from VizDoom environment and agent can be trained with globally consistent state representation with respect to dense depth maps.

Acknowledgement. The work was supported by the Russian Science Foundation under grant 22-11-00323 and performed at National Research University Higher School of Economics, Moscow, Russia.

References

1. Anderson, P., et al.: Vision-and-language navigation: interpreting visually-grounded navigation instructions in real environments. In: Proceedings of CVPR, pp. 3674–3683 (2018)
2. Badia, A.P., et al.: Agent57: outperforming the Atari human benchmark. In: International Conference on Machine Learning, pp. 507–517. PMLR (2020)
3. Bakhanova, M., Makarov, I.: Deep reinforcement learning in VizDoom via DQN and actor-critic agents. In: Rojas, I., Joya, G., Català, A. (eds.) IWANN 2021. LNCS, vol. 12861, pp. 138–150. Springer, Cham (2021). https://doi.org/10.1007/978-3-030-85030-2_12

4. Beattie, C., et al.: DeepMind Lab. arXiv arXiv:1612.03801 (2016)
5. Bellemare, M.G., Naddaf, Y., Veness, J., Bowling, M.: The arcade learning environment: an evaluation platform for general agents. J. Artif. Intell. Res. **47**, 253–279 (2013)
6. Bilkhu, M., Wang, S., Dobhal, T.: Attention is all you need for videos: self-attention based video summarization using universal transformers. arXiv arXiv:1906.02792 (2019)
7. Chen, L., et al.: Decision transformer: reinforcement learning via sequence modeling. arXiv preprint arXiv:2106.01345 (2021)
8. Chen, M.X., et al.: The best of both worlds: combining recent advances in neural machine translation. arXiv arXiv:1804.09849 (2018)
9. Chung, J., Gulcehre, C., Cho, K., Bengio, Y.: Empirical evaluation of gated recurrent neural networks on sequence modeling. arXiv arXiv:1412.3555 (2014)
10. Degris, T., Pilarski, P.M., Sutton, R.S.: Model-free reinforcement learning with continuous action in practice. In: 2012 American Control Conference (ACC), pp. 2177–2182. IEEE (2012)
11. Devlin, J., Chang, M.W., Lee, K., Toutanova, K.: Bert: pre-training of deep bidirectional transformers for language understanding. arXiv arXiv:1810.04805 (2018)
12. Dosovitskiy, A., et al.: An image is worth 16 × 16 words: transformers for image recognition at scale. arXiv arXiv:2010.11929 (2020)
13. Ha, D., Schmidhuber, J.: World models. arXiv arXiv:1803.10122 (2018)
14. Hu, S., Zhu, F., Chang, X., Liang, X.: UPDeT: universal multi-agent reinforcement learning via policy decoupling with transformers. arXiv arXiv:2101.08001 (2021)
15. HuggingFace.co: Vision transformers model implementation, transformers Python package. https://huggingface.co/transformers/model_doc/vit.html
16. Kaelbling, L.P., Littman, M.L., Cassandra, A.R.: Planning and acting in partially observable stochastic domains. Artif. Intell. **101**(1–2), 99–134 (1998)
17. Kamaldinov, I., Makarov, I.: Deep reinforcement learning in match-3 game. In: Proceedings of CoG, pp. 1–4. IEEE (2019)
18. Kapturowski, S., Ostrovski, G., Quan, J., Munos, R., Dabney, W.: Recurrent experience replay in distributed reinforcement learning. In: International Conference on Learning Representations (2018)
19. Kar, P.: A2C, ACKTR and A2T implementations for ViZDoom. https://github.com/p-kar/a2c-acktr-vizdoom
20. Kempka, M., Wydmuch, M., Runc, G., Toczek, J., Jaśkowski, W.: ViZDoom: a doom-based ai research platform for visual reinforcement learning. In: 2016 IEEE Conference on Computational Intelligence and Games (CIG), pp. 1–8. IEEE (2016)
21. Korinevskaya, A., Makarov, I.: Fast depth map super-resolution using deep neural network. In: 2018 IEEE International Symposium on Mixed and Augmented Reality Adjunct (ISMAR-Adjunct), pp. 117–122. IEEE (2018)
22. Li, J., Wu, Y., Gaur, Y., Wang, C., Zhao, R., Liu, S.: On the comparison of popular end-to-end models for large scale speech recognition. arXiv arXiv:2005.14327 (2020)
23. Makarov, I., Aliev, V., Gerasimova, O.: Semi-dense depth interpolation using deep convolutional neural networks. In: Proceedings of the 2017 ACM on Multimedia Conference, pp. 1407–1415. ACM, NY, USA (2017)
24. Makarov, I., Aliev, V., Gerasimova, O., Polyakov, P.: Depth map interpolation using perceptual loss. In: 2017 IEEE International Symposium on Mixed and Augmented Reality (ISMAR-Adjunct), pp. 93–94. IEEE, NY, USA (2017)

25. Makarov, I., Borisenko, G.: Depth inpainting via vision transformer. In: 2021 IEEE International Symposium on Mixed and Augmented Reality Adjunct (ISMAR-Adjunct), pp. 286–291. IEEE (2021)

26. Makarov, I., Kashin, A., Korinevskaya, A.: Learning to play pong video game via deep reinforcement learning. In: AIST (Supplement), pp. 236–241 (2017)

27. Makarov, I., Polyakov, P.: Smoothing Voronoi-based path with minimized length and visibility using composite Bezier curves. In: AIST (Supplement), pp. 191–202 (2016)

28. Makarov, I., Polyakov, P., Karpichev, R.: Voronoi-based path planning based on visibility and kill/death ratio tactical component. In: Proceedings of AIST (2018)

29. Makarov, I., Savostyanov, D., Litvyakov, B., Ignatov, D.I.: Predicting winning team and probabilistic ratings in "Dota 2" and "Counter-Strike: Global Offensive" video games. In: van der Aalst, W.M.P., et al. (eds.) AIST 2017. LNCS, vol. 10716, pp. 183–196. Springer, Cham (2018). https://doi.org/10.1007/978-3-319-73013-4_17

30. Makarov, I., et al.: First-person shooter game for virtual reality headset with advanced multi-agent intelligent system. In: Proceedings of the 24th ACM International Conference on Multimedia, pp. 735–736 (2016)

31. Makarov, I., Tokmakov, M., Tokmakova, L.: Imitation of human behavior in 3d-shooter game. In: Analysis of Images, Social Networks and Texts, AIST 2015, p. 64 (2015)

32. Makarov, I., et al.: Modelling human-like behavior through reward-based approach in a first-person shooter game. In: EEML Proceedings (2016)

33. Maslov, D., Makarov, I.: Online supervised attention-based recurrent depth estimation from monocular video. PeerJ Comput. Sci. **6**, e317 (2020)

34. Merkx, D., Frank, S.L.: Human sentence processing: Recurrence or attention? arXiv arXiv:2005.09471 (2020)

35. Mnih, V., et al.: Asynchronous methods for deep reinforcement learning. In: International Conference on Machine Learning, pp. 1928–1937. PMLR (2016)

36. Mnih, V., et al.: Human-level control through deep reinforcement learning. nature **518**(7540), 529–533 (2015)

37. Parisotto, E., et al.: Stabilizing transformers for reinforcement learning. In: ICML, pp. 7487–7498. PMLR (2020)

38. Pathak, D., Agrawal, P., Efros, A.A., Darrell, T.: Curiosity-driven exploration by self-supervised prediction. In: International Conference on Machine Learning, pp. 2778–2787. PMLR (2017)

39. Paul, S., Chen, P.Y.: Vision transformers are robust learners. arXiv arXiv:2105.07581 (2021)

40. Samvelyan, M., et al.: The starcraft multi-agent challenge. arXiv arXiv:1902.04043 (1902)

41. Schrittwieser, J., et al.: Mastering Atari, Go, chess and shogi by planning with a learned model. nature **588**(7839), 604–609 (2020)

42. Tuli, S., Dasgupta, I., Grant, E., Griffiths, T.L.: Are convolutional neural networks or transformers more like human vision? arXiv:2105.07197 (2021)

43. Upadhyay, U., Shah, N., Ravikanti, S., Medhe, M.: Transformer based reinforcement learning for games. arXiv preprint arXiv:1912.03918 (2019)

44. Vaswani, A., et al.: Attention is all you need. In: Advances in Neural Information Processing Systems, pp. 5998–6008 (2017)

45. Wang, Y., et al.: End-to-end video instance segmentation with transformers. In: Proceedings of the IEEE/CVF Conference on Computer Vision and Pattern Recognition, pp. 8741–8750 (2021)

46. Watkins, C.J., Dayan, P.: Q-learning. Mach. Learn. **8**(3–4), 279–292 (1992)
47. Wu, Y., Wu, Y., Gkioxari, G., Tian, Y.: Building generalizable agents with a realistic and rich 3d environment. arXiv arXiv:1801.02209 (2018)
48. Wu, Y., Mansimov, E., Liao, S., Radford, A., Schulman, J.: OpenAI baselines: ACKTR & A2C (2017). https://openai.com/blog/baselines-acktr-a2c
49. Wu, Y., Tian, Y.: Training agent for first-person shooter game with actor-critic curriculum learning (2016). https://openreview.net/forum?id=Hk3mPK5gg
50. Xu, Y., Chen, L., Fang, M., Wang, Y., Zhang, C.: Deep reinforcement learning with transformers for text adventure games. In: 2020 IEEE Conference on Games (CoG), pp. 65–72. IEEE (2020)
51. Zhu, Y., et al.: Target-driven visual navigation in indoor scenes using deep reinforcement learning. In: 2017 IEEE International Conference on Robotics and Automation (ICRA), pp. 3357–3364. IEEE (2017)

Text CAPTCHA Traversal via Knowledge Distillation of Convolutional Neural Networks: Exploring the Impact of Color Channels Selection

Valery Terekhov(D), Valery Chernenky, and Denis Ishkov$^{(\boxtimes)}$(D)

Bauman Moscow State Technical University, 105005 Moscow, Russia
terekchow@bmstu.ru, defasium@yandex.ru

Abstract. While most of the existing works have investigated the recognition of a fixed-length CAPTCHA, the authors of the article propose to apply knowledge distillation to approximate the predictions of recurrent convolutional neural networks. Such models have proven themselves well in predicting the dynamic length of characters in images. The paper studies the influence of the size and complexity of the training sample on the quality of recognition. The authors studied the effect of individual color channels and their linear combination on the final quality of models. An estimate of the importance of each color channel was acquired using the trainable scalar coefficients in a linear combination. The results obtained made it possible to reduce the size of the input data for recognition without loss in quality of recognition, as well as speed up the training of the model. The analysis of model errors allowed us to offer suggestions for improving ways of countering automatic recognition.

Keywords: Text CAPTCHA recognition · Convolutional neural network · Deep learning · Knowledge distillation

1 Introduction

Nowadays, the number of automatic Internet programs or bots that allow performing diverse actions without human participation is growing everywhere: from data collection to communication in social networks [10]. An excessive number of bots running on the site can slow down its work or lead to failure due to too many requests in a short period of time. To avoid such a scenario, various means of protection and detection of abnormal users of a Web resource were created. One of the very first methods of protection is Completely Automated Public Turing test to tell Computers and Humans Apart (CAPTCHA) [18]. Initially looking like a distorted text on an image, the CAPTCHA has undergone many changes and today can be provided in many forms. With the increase in the number of various protective mechanisms against bots, there has also been an increase in ways to bypass such protection. The result of this confrontation is

E. Burnaev et al. (Eds.): AIST 2021, CCIS 1573, pp. 111–122, 2022.
https://doi.org/10.1007/978-3-031-15168-2_10

an increasingly sophisticated and stable CAPTCHA and increasingly complex methods of its recognition. However, the original text form is still the most common and is used by many widespread sites, e.g. Yandex, VKontakte, Steam, WebMoney, Wikipedia, etc. In the paper of 2018 [21] to bypass the CAPTCHA of well-known Internet companies the authors used various solutions, in particular, paid automatic recognition services. Such services distribute requests among labeling experts. This approach has obvious advantages such as: adaptability to changing forms and recognition of characters that are not fixed in length and alphabet. There are also disadvantages, namely, a long response and not perfect recognition due to the presence of a human factor.

As an alternative to paid services, deep learning methods can be used in order to bypass the protection automatically [19,23]. However, with the growth of the quality of CAPTCHA recognition, the requirement for the size of the training sample and the complexity of models increases.

Many solutions have been proposed to address the first problem. The most promising direction is the generation of synthetic CAPTCHA [12,20] from the initial set, which does not exceed about a thousand images. Along with this, it was shown that it is possible to apply transfer learning from a similar domain [11] and active learning [16].

The problem of recognizing a dynamic number of characters is solved either by preliminary segmentation of the image [9] or by sequential recognition using attached labels [17]. Such approaches require recognizing an image several times, so they are not optimal. It's important to note that in some cases the segmentation approach appears to be the most feasible one, e.g. in the case of spatial determining of tree crowns [3].

The problem of the size of models was practically not touched upon earlier in the field of CAPTCHA recognition. It is proposed to solve this problem using Knowledge distillation [8] - which is not so much a technique for compressing neural networks, as using large trained models to train smaller models.

The problem of a poor variety of colors in CAPTCHA images may lead to a reduction of training examples required to achieve the desired recognition quality. Insufficient diversity of colors explained by only one color channel can result in faster training of deep learning models on this certain subset of image features. Thus the ability to estimate the influence of each channel can identify the need for reconsidering the CAPTCHA color scheme.

The paper proposes a method for training lightweight convolutional neural networks without recurrent layers for recognizing text CAPTCHA of various lengths using knowledge distillation. The importance of channels was acquired via implementing the custom neural network layer which learns the linear combination of color channels. This can give an understanding which is necessary to build more robust CAPTCHA designs against automatic recognition methods. For experiments, CAPTCHA images from the VKontakte social network are used. Recommendations were formulated based on the acquired results.

2 Methodology

2.1 Problem Formulation and Proposed Method Description

Let there be a test set S' that does not intersect with the training set S, which consists of pairs (x, y), where x is the input tensor of an image with a text CAPTCHA, and y is its character labels. In this paper, we are interested in the Label Error Rate (LER) of the classifier h as the normalized Levenshtein distance between its predictions and the true values on the set S' as an expression (1):

$$LER(h, S') = \frac{1}{H} \sum_{(x,y) \in S'} ED(h(x), y) \tag{1}$$

where H is the total number of target labels in S', $ED(p, q)$ is the Levenshtein distance between two sequences p and q, i.e. the minimum number of inserts, substitutions and deletions required to change p to q. Unfortunately, this metric is not differentiable and cannot be used as a loss function. The loss function proposed in [5] called connectionist temporal classification (CTC) is derived from the maximum likelihood principle. The goal of maximum likelihood training is to simultaneously maximize the logarithmic probabilities p of all correct classifications z in S. Which means minimizing the following objective function (2):

$$L_{CTC} = - \sum_{(x,z) \in S} \ln(p(z \mid x)) \tag{2}$$

With the help of "dynamic programming", the problem of finding all possible paths of z can be represented as a recursive convergent sum. Thus, the derivative [5] can be taken from the expression (2), which allows using it as a loss function.

To train lightweight models, the authors of the article propose to apply the knowledge distillation. It transfers knowledge from one deep learning model (teacher) to another (student). The additional term in the loss function, originally proposed in the article [8], minimizes the Kullback-Leibler divergence KL between the predictions of the teacher and the student. The full loss function is as follows:

$$L_{KD} = (1 - \lambda) \times L_{CTC} + \lambda \times \rho^2 \times KL(\sigma(\frac{z_T}{\rho}) \mid \sigma(\frac{z_S}{\rho})) \tag{3}$$

where ρ is the temperature, λ is the weighting factor controlling the trade-off between the two losses, σ is the softmax function, z_T and z_S are predictions of the output layer before applying the activation of the teacher and student, respectively.

When knowledge distillation is used on a labeled dataset, true labels are used for calculating L_{CTC}. When it is used on an unlabelled dataset, the teacher's predictions are used. There may be errors in it, that's why in the following sections we will call such inaccurate labeling *noisy*.

The Kullback-Leibler divergence is an asymmetric function of the arguments, which leads to confusion in the implementation. To avoid this, the authors of the article will use the Jenson-Shannon divergence instead in their experiments.

2.2 Data Gathering and Analysis

The dataset for training models is a set of collected text CAPTCHA of the social network VKontakte. This CAPTCHA is presented in two types, "normal" and "hard", while the unique number of characters used in both is limited to 21 (see Fig. 1). The characters include English letters, as well as Arabic digits. Both variants have a resolution of 130 by 50 pixels, distorted elongated characters and several overlapping lines. The "normal" set has from 4 to 5 characters with 2 overlapping lines. The "hard" one contains from 6 to 7 characters and 4 overlapping lines.

Two methods were utilized in order to acquire the data. Initially, the unlabelled sets for "normal" and "hard" type of CAPTCHA were parsed using a program written by the authors. Thus, sets of 1.5 and 2.0 million training examples were collected. The resolution of the sides of each obtained image was reduced to the nearest multiples of the power of two, namely 64 and 128, respectively. The resulting examples were partially labeled manually to obtain the first iteration models.

After obtaining recognition models of acceptable quality, they were used for automatic labeling by means of the Selenium WebDriver library. Human participation in this process is not necessary due to the presence of feedback from the validator resource (the social network VKontakte) after sending the expected result. If the characters are correctly recognized from the image, the on-screen verification form disappears. Otherwise, the screen form remains, and the text CAPTCHA is replaced.

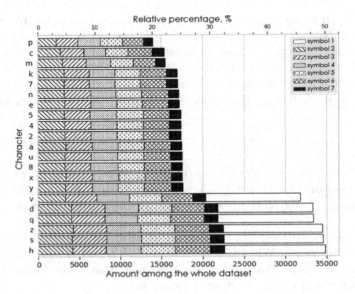

Fig. 1. Distribution of characters in the collected "hard" dataset containing from 6 to 7 characters in the image.

After expanding the training sample in this way, the authors of the article propose to retrain the model on new data. It is worth noting that there is a daily limit on the same type of actions for user profiles. Thus, the permissible number of text CAPTCHA from one account per day, according to rough estimates, cannot exceed 3 thousand images.

2.3   Training of the Classifier Model

The training of the classifier model is proposed to be carried out in an iterative fashion. In this case, one iteration means collecting an additional dataset, combining it with the original one and training new model on the expanded dataset. The weights of the model are randomly initialized, and the previous values obtained in the last iteration are not used. This allows us to explore other local minima. Training images are augmented by various distortions, color correction, lossy compression, blurring and adding noise. During training, dropout layers [15] are used, which act as regularization. On small sets, a more aggressive regularization is required. At the beginning of the training, the warm up technique of the learning rate up to 5e−4 is used [4]. The gradient descent algorithm uses RMSProp [7] without momentum, the size of the minibatch during training is 128. The dataset is divided into training and test splits, where the latter is 10% of the original set. The quality of the model is evaluated using the previously described metric LER (1) on the test set. If there is no improvement in this metric for 10 epochs, the training step is reduced by an order of magnitude.

To train a large model or teacher, the CRNN architecture is used (see Fig. 2). The CNN backbone uses architecture similar to GoogLeNet, which is often used for OCR tasks [22]. It contains 5.5 million parameters for training. The recurrent part consists of two blocks of bidirectional LSTM layers [6]. This type of recurrent layer is widely used for various sequential data, e.g. in epidemiological forecasting [2]. The last layer is fully connected, with the softmax activation function.

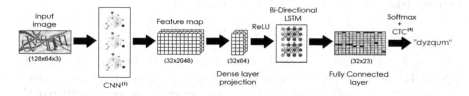

Fig. 2. Structure of the teacher model.

To train a student model, MobileNetV2 [14] is used for the "hard" and ShuffleNetV2 [13] for the "normal" datasets. The student contains 260 thousand trained parameters. Instead of a recurrent part, one-dimensional convolutional layers with kernel sizes of 3 and 5 with subsequent concatenation are used, 128 filters in total. Otherwise, the architecture of the student does not differ from the teacher.

The CTC loss function is used as the teacher's loss function. It is proposed to use two different loss functions for teaching the student: CTC is used to obtain the baseline model before knowledge distillation, while a weighted sum with a coefficient $\lambda = 20\%$ and a temperature $\rho = 2$ is used for distillation (3). A student is trained after achieving the necessary quality of a teacher. Then it labels the millions of images collected earlier, and the resulting noisy predictions are used for training. The use of an extended noisy dataset is due to the fact that a small set may not be enough to distill knowledge [1]. After training on the extended set, the model is further fine-tuned on the original dataset with the CTC loss function (2). All parameters, except for the parameters of the last fully connected layer, are fixed and are not updated.

2.4   Estimating the Contribution of Color Channels

In order to grasp the importance of RGB channels on the accuracy of the student model, we can train it 5 times on the "normal" dataset with CTC loss function (2) on each of the color channels and evaluate the quality on the hold-out set. Each run has a limited 3 h of GPU budget. For averaging the results, we use the harmonic mean of the recognition error. Harmonic mean has a strong bias toward smaller values in the sample, meaning that the effect of large outliers on the resulting value will be much smaller than the effect of small outliers. This method can answer which one of the color channels has the greatest impact on the model generalization ability but can't answer how they relate with each other. To do so we can also train models on pair of channels, e.g. red and blue. To compare obtained results we can investigate other color projection methods, like training on value channel from HSV color representation or training on grayscale images. There are several color-to-grayscale conversions, like taking the arithmetic mean of color pixels or averaging maximum and minimum values of a pixel. However, in our experiments we will use the most common algorithm for conversion proposed in the NTSC standard (4):

$$H(R, G, B) = 0.299 \times R + 0.587 \times G + 0.114 \times B \tag{4}$$

where R, G, and B are corresponding to red, green, and blue color channels of a CAPTCHA image respectively and H is a mapping function to grayscale space.

This formula closely represents the average person's relative perception of the brightness of red, green, and blue light but for the case of the certain dataset may not be the best linear combination of colors. In order to find the optimal linear combination, the authors of the paper implemented the custom neural network layer with 3 learnable parameters which receive RGB image as input and outputs image with one color channel. This operation is equivalent to applying one convolutional two-dimensional filter with a kernel size of 1 without bias parameter. Produced channel from such custom layer is in fact the linear combination of color channels. Parameters are initialized with NTSC coefficients and are updated like all other neural network weights during training and would converge to the optimal values. Thus, by taking the magnitude of learned parameters we can estimate the importance of each color channel on a specific dataset.

It's worth mentioning that training a model on one channel instead of three is faster and requires less GPU memory for the same batch. The inference time will expected to be faster. However, we spend some computational time to merge channels, so in an ideal case scenario, we would like to train our model on one of the initial RGB channels due to the fact that indexing an array is usually a less costly operation.

3 Experiments

To get an estimate of the quality of the model based on real data, we used feedback from the VKontakte social network as was discussed earlier. Each instance contained a timestamp, recognized text, and a success flag. With the help of bootstrapping, the expectation of quality was estimated by time intervals. When measuring, NVIDIA Tesla P100 PCIe 16 GB was used as a graphics computing device (GPU), and a single-core Intel Xeon processor (CPU) with a clock frequency of 2.3 GHz was used.

The analysis of the results of the experiment allows us to conclude that on the "normal" dataset of 40 thousand images, a teacher model reaches a quality of about 99% (see Fig. 3). Quality refers to the fraction of correctly recognized CAPTCHA. On the "hard" set of 70 thousand images, the same model with the same number of parameters reaches about 98% accuracy. Student models lose to the teacher by 10 and 20%, respectively. The "+" sign next to the name of the dataset indicates that the model is being trained on an extended *noisy* set.

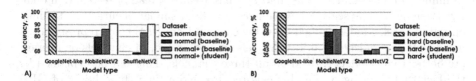

Fig. 3. Achieved recognition quality by different models: A) the result on a "normal" dataset; B) the result on a "hard" dataset.

Based on the obtained results, it can be concluded that CAPTCHA protection measures, such as an increase in the number of characters and a more diverse background, have a negligible effect on the recognition quality if there are sufficient examples for training and a large capacity of the model complexity itself. At the same time, for models with an insufficient number of parameters, increasing the number of characters makes the recognition task less feasible. The size of the training sample affects the final quality of the model, which is not surprising. The authors of the paper investigated this dependency. A hypothesis about the form of this dependence was formulated in (5):

$$f(x) = \frac{a}{1 - e^{-k \times \ln(\frac{x}{x_0})}} + b \tag{5}$$

where a is the scaling factor, b is the bias, k is the growth rate of the function in the linearity zone, x_0 is the offset along the x-axis. This function belongs to the family of the sigmoid class functions and contains two saturation zones and one linear zone.

The choice of such function is confirmed by the fact that if there is not enough data, the model can't generalize at all. Its quality then monotonically increases as more data is given. This process continues until the moment when the complexity of the model does not allow achieving better generalization ability. The function proposed above fully corresponds to this phenomenon. To obtain the required values, the teacher was trained on subsamples of data sets three times, after which the best model was selected to estimate the quality of recognition with bootstrap. Then, using the obtained values, the coefficients of the function were found using the least-squares method (OLS). The final form of the dependence of the teacher's quality on the number of examples in the training sample can be seen in Fig. 4. The results allow us to make sure that the "hard" dataset requires more training examples - the saturation zone occurs in the region of 100 thousand images against 20 thousand for the "normal" dataset. From the data obtained, we can also conclude that the model begins to generalize to unseen images after training on 200–400 examples. Thus, in order to get a first CAPTCHA recognition model, it is enough to manually label several hundred images which can take only a few hours.

The impact of color channels on the model accuracy trained on the "normal" dataset can be seen in Fig. 5. We can conclude from the obtained results that training on red channel compared green and blue gives on average better results. Training on blue channel is unstable as it has greater spread of recognition quality. Training on grayscale NTSC images results in similar accuracy as training on red channel. Model trained on value channel from HSV color representation has decent generalization ability. Models trained on pair of channels on average are better than models trained on all three channels. This can be explained by the fact that training on all channels requires more than three hours of GPU budget to converge. Among all results training with a learnable weighted sum of color channels (MIX) is better than models trained on raw RGB images. The authors of the paper conducted Mann-Whitney U test and obtained statistically significant results with a p-value of 0.00786. The acquired parameters which are summarized in Table 1 can lead us to the conclusion that the red channel indeed has the greatest impact on the given CAPTCHA dataset. The cause for that may be the fact that all images presented in datasets are in cool colors with a predominance of blue shades. In that case, the best strategy is to give more attention to colors of the opposite spectrum. We train the student model on the noisy "normal" dataset on the red channel and obtained an increase in the recognition accuracy by 1% compared to training on full channels. We also investigated the ways to eliminate this improvement by introducing more color diversity by randomly changing the hue of each image before feeding it to the

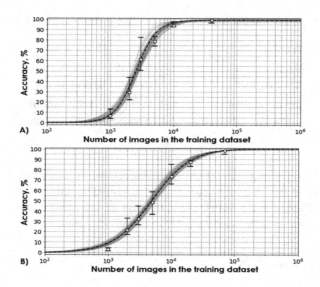

Fig. 4. The dependence of the recognition quality of the teacher model on the size of the training sample: A) the result on the "normal" dataset; B) the result on the "hard" dataset.

neural network. We obtained somewhat uniform values for parameters with a small bias towards the blue channel as can be seen in Table 1. Thus, by training on one of the channels on modified images we can expect to acquire at best the similar quality to training on the green channel on original images.

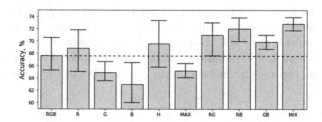

Fig. 5. The dependence of the recognition quality of the student model on the color channels used for training. Error bars indicate 95% confidence intervals. H corresponds to NTSC grayscale, MAX corresponds to the value channel in HSV color representation and MIX corresponds to a learnable linear combination of color channels.

During the study, the errors of the teacher model were analyzed and divided into two global categories of errors: an incorrectly recognized character and an incorrectly recognized length. The error analysis showed that 60% of the teacher's errors are due to the incorrect CAPTCHA length. When identical characters are arranged sequentially, the model recognizes them as a single character. If we talk

Table 1. Estimated learnable coefficients in the linear combination of color channels.

Variable	R	G	B
Mean, original	**0.5739**	0.3596	0.0664
Std, original	0.0179	0.0090	0.0102
Mean, random hue changed	0.3170	0.3171	**0.3658**
Std, random hue changed	0.0044	0.0071	0.0030

about errors in replacing characters, the model confuses similar characters. This can be explained by the invariant nature of convolutional filters used in the architecture. Here are the most popular pairs of characters that confuse the classifier: "c" and "e", "v" and "y", "h" and "n", "c" and "q", "n" and "u".

4   Conclusion

The authors of the work proposed a method for training models for recognizing text CAPTCHA of variable length with a volume of several megabytes. The proposed iterative semi-automatic training with feedback from a validator resource is able to obtain a recognition model in a short time. In this case, you need to label only a small number of examples manually before the first iteration. The small size of the model and the absence of recurrent layers allowed us to achieve acceptable results in the speed of the neural network, without losing the high quality of recognition. The improvement of the quality of the student model was carried out using the knowledge distillation technique. The resulting increase after distillation of 5% indicates the competitiveness of the proposed method. The authors showed the importance of color diversity in CAPTCHA design. Otherwise, there is a vulnerability by training on the most informative color channel. This lowers the GPU memory consumption without loss in quality of recognition, as well as speeds up the training of the model. The performed analysis of classifier errors is intended to improve the ways of countering automatic recognition.

References

1. Ba, L.J., Caruana, R.: Do deep nets really need to be deep? arXiv preprint arXiv:1312.6184 (2013)
2. Burdakov, A.V., Ukharov, A.O., Myalkin, M.P., Terekhov, V.I.: Forecasting of Influenza-like illness incidence in Amur Region with neural networks. In: Kryzhanovsky, B., Dunin-Barkowski, W., Redko, V., Tiumentsev, Y. (eds.) NEUROINFORMATICS 2018. SCI, vol. 799, pp. 307–314. Springer, Cham (2019). https://doi.org/10.1007/978-3-030-01328-8_37

3. Eroshenkova, D.A., Terekhov, V.I., Khusnetdinov, D.R., Chumachenko, S.I.: Automated determination of forest-vegetation characteristics with the use of a neural network of deep learning. In: Kryzhanovsky, B., Dunin-Barkowski, W., Redko, V., Tiumentsev, Y. (eds.) NEUROINFORMATICS 2019. SCI, vol. 856, pp. 295–302. Springer, Cham (2020). https://doi.org/10.1007/978-3-030-30425-6_34
4. Goyal, P., et al.: Accurate, large minibatch SGD: training ImageNet in 1 hour. arXiv preprint arXiv:1706.02677 (2017)
5. Graves, A., Fernández, S., Gomez, F., Schmidhuber, J.: Connectionist temporal classification: labelling unsegmented sequence data with recurrent neural networks. In: Proceedings of the 23rd International Conference on Machine Learning, pp. 369–376 (2006)
6. Graves, A., Schmidhuber, J.: Framewise phoneme classification with bidirectional LSTM and other neural network architectures. Neural Netw. **18**(5–6), 602–610 (2005)
7. Hinton, G., Srivastava, N., Swersky, K.: Neural networks for machine learning lecture 6a overview of mini-batch gradient descent. Cited on **14**(8), 2 (2012)
8. Hinton, G., Vinyals, O., Dean, J.: Distilling the knowledge in a neural network. arXiv preprint arXiv:1503.02531 (2015)
9. Hussain, R., Gao, H., Shaikh, R.A.: Segmentation of connected characters in text-based captchas for intelligent character recognition. Multimedia Tools Appl. **76**(24), 25547–25561 (2017)
10. Klopfenstein, L.C., Delpriori, S., Malatini, S., Bogliolo, A.: The rise of bots: a survey of conversational interfaces, patterns, and paradigms. In: Proceedings of the 2017 Conference on Designing Interactive Systems, pp. 555–565 (2017)
11. Kushchuk, D.O., Ryndin, M.A., Yatskov, A.K., Varlamov, M.I.: Using domain adversarial learning for text captchas recognition. Proc. Inst. Syst. Program. RAS **32**(4), 203–216 (2020)
12. Li, C., Chen, X., Wang, H., Wang, P., Zhang, Y., Wang, W.: End-to-end attack on text-based captchas based on cycle-consistent generative adversarial network. Neurocomputing **433**, 223–236 (2021)
13. Ma, N., Zhang, X., Zheng, H.-T., Sun, J.: ShuffleNet V2: practical guidelines for efficient CNN architecture design. In: Ferrari, V., Hebert, M., Sminchisescu, C., Weiss, Y. (eds.) Computer Vision – ECCV 2018. LNCS, vol. 11218, pp. 122–138. Springer, Cham (2018). https://doi.org/10.1007/978-3-030-01264-9_8
14. Sandler, M., Howard, A., Zhu, M., Zhmoginov, A., Chen, L.C.: MobileNetV2: inverted residuals and linear bottlenecks. In: Proceedings of the IEEE Conference on Computer Vision and Pattern Recognition, pp. 4510–4520 (2018)
15. Srivastava, N., Hinton, G., Krizhevsky, A., Sutskever, I., Salakhutdinov, R.: Dropout: a simple way to prevent neural networks from overfitting. J. Mach. Learn. Res. **15**(1), 1929–1958 (2014)
16. Stark, F., Hazırbas, C., Triebel, R., Cremers, D.: Captcha recognition with active deep learning. In: Workshop New Challenges in Neural Computation, vol. 2015, p. 94. Citeseer (2015)
17. Thobhani, A., Gao, M., Hawbani, A., Ali, S.T.M., Abdussalam, A.: Captcha recognition using deep learning with attached binary images. Electronics **9**(9), 1522 (2020)
18. von Ahn, L., Blum, M., Hopper, N.J., Langford, J.: CAPTCHA: using hard AI problems for security. In: Biham, E. (ed.) EUROCRYPT 2003. LNCS, vol. 2656, pp. 294–311. Springer, Heidelberg (2003). https://doi.org/10.1007/3-540-39200-9_18

19. Wang, J., Qin, J.H., Xiang, X.Y., Tan, Y., Pan, N.: Captcha recognition based on deep convolutional neural network. Math. Biosci. Eng. **16**(5), 5851–5861 (2019)
20. Ye, G., et al.: Yet another text captcha solver: a generative adversarial network based approach. In: Proceedings of the 2018 ACM SIGSAC Conference on Computer and Communications Security, pp. 332–348 (2018)
21. Zhao, B., et al.: Towards evaluating the security of real-world deployed image captchas. In: Proceedings of the 11th ACM Workshop on Artificial Intelligence and Security, pp. 85–96 (2018)
22. Zhong, Z., Jin, L., Xie, Z.: High performance offline handwritten Chinese character recognition using GoogleNet and directional feature maps. In: 2015 13th International Conference on Document Analysis and Recognition (ICDAR), pp. 846–850. IEEE (2015)
23. Zi, Y., Gao, H., Cheng, Z., Liu, Y.: An end-to-end attack on text captchas. IEEE Trans. Inf. Forensics Secur. **15**, 753–766 (2019)

Data Analysis and Machine Learning

Dropout Strikes Back: Improved Uncertainty Estimation via Diversity Sampling

Kirill Fedyanin[1(✉)], Evgenii Tsymbalov[1,2(✉)], and Maxim Panov[1(✉)]

[1] Skolkovo Institute of Science and Technology (Skoltech), Moscow, Russia
{k.fedyanin,m.panov}@skoltech.ru
[2] Yandex, Moscow, Russia
etsymba@yandex-team.ru

Abstract. Uncertainty estimation for machine learning models is of high importance in many scenarios such as constructing the confidence intervals for model predictions and detection of out-of-distribution or adversarially generated points. In this work, we show that modifying the sampling distributions for dropout layers in neural networks improves the quality of uncertainty estimation. Our main idea consists of two main steps: computing data-driven correlations between neurons and generating samples, which include maximally diverse neurons. In a series of experiments on simulated and real-world data, we demonstrate that the diversification via *determinantal point processes*-based sampling achieves state-of-the-art results in uncertainty estimation for regression and classification tasks. An important feature of our approach is that it does not require any modification to the models or training procedures, allowing straightforward application to any deep learning model with dropout layers.

Keywords: Uncertainty estimation · Neural networks · Dropout · Determinantal point processes

1 Introduction

Uncertainty estimation (UE) recently became a very active area of research in deep learning. Neural networks usually are treated as black boxes, and in general, they are prone to overconfidence [9,15]. Uncertainty estimation methods aim to help overcome this drawback by identifying potentially erroneous predictions. This can be especially important for error-critical applications like medical diagnostics [4] or autonomous car driving [8]. Another important application for uncertainty estimation is active learning [34]. The majority of sampling criteria in active learning are based on estimates of uncertainty, which makes it important to obtain high-quality uncertainty estimates.

There are several main approaches for uncertainty estimation for deep neural networks. Bayesian neural networks (BNN) and variational inference in particular represent a natural way for uncertainty estimation due to availability of

© The Author(s), under exclusive license to Springer Nature Switzerland AG 2022
E. Burnaev et al. (Eds.): AIST 2021, CCIS 1573, pp. 125–137, 2022.
https://doi.org/10.1007/978-3-031-15168-2_11

well-defined posteriors, but they can be prohibitively slow for large-scale applications. The usage of dropout at the inference stage was shown to be good and efficient approximation to BNNs [9,10]. The ensembles of independently trained models [25] have state-of-the-art performance in many tasks requiring uncertainty estimation [37]. Recently, forcing models in ensembles to be more diverse was shown to improve results even further [22]. The drawback of ensembles is that we need to train and use multiple models that require additional resources, i.e. more memory to store models and more computing power for training.

In this work, we aim to develop a new approach for dropout-based uncertainty estimation. Usually there are many highly correlated neurons in neural networks, which results in a slow convergence of estimates based on the standard uniform sampling in dropout layers. We propose to estimate correlations between neurons based on the data and sample the most diverse neurons in order to improve the convergence of the estimates and, as a result, the quality of uncertainty estimates. As a particular realization of the general idea, we suggest sampling dropout masks using the machinery of determinantal point processes (DPP) [24] which are known to give diverse samples.

We summarize the main contributions of the paper as follows:

- We propose two DPP-based sampling methods for neural networks with dropout. Our approach requires to train only a single model and adds only small overhead on the inference stage compared to plain MC dropout.
- We compare different dropout-based approaches for uncertainty estimation in an extensive series of experiments for real-world regression and classification datasets. The results show superior performance of proposed DPP-based approaches.
- Importantly, the proposed methods show high quality of uncertainty estimation even for very small number of stochastic passes through the network, thus opening the possibility to significantly speed up the inference stage.

The rest of the paper is organized as follows. Section 2 introduces the proposed method for DPP-based sampling from neural networks with dropout. In Sect. 3, we show the efficiency of the proposed approach in the problem of uncertainty estimation. Section 4 gives an overview of the related work on uncertainty estimation for neural networks. Section 5 concludes the study and highlights some directions for future work.

2 Methods

2.1 Neural Networks with Dropout as Implicit Ensembles

We start by considering a standard fully connected layer in a neural network

$$S_i^h = \sum_{j=1}^{N_{h-1}} w_{ij}^h O_j^{h-1}, \quad i = 1, \ldots, N_h, \tag{1}$$

where $O_i^h = \sigma(S_i^h)$ is an output of the h-th layer of the neural network given by a non-linear transformation $\sigma(\cdot)$ of the corresponding pre-activation S_i^h.

An application of dropout to neurons results in the following formula for the pre-activations:

$$S_i^h = \sum_{j=1}^{N_{h-1}} \frac{1}{1-p} m_j^h w_{ij}^h O_j^{h-1}, \quad i = 1, \ldots, N_h, \tag{2}$$

where m_j^h are Bernoulli random variables with a probability of 0 equal to p. The outputs O_i^h of the h-th layer remain to be computed by $O_i^h = \sigma(S_i^h)$. Note that if an input variable of neural network is denoted by $\mathbf{x}$, then output of every layer is a function of $\mathbf{x}$, i.e., $O_i^h = O_i^h(\mathbf{x})$.

Let us denote the vector of dropout weights m_j^h for the h-th layer by $\mathbf{m}_h = (m_1^h, \ldots, m_{N_h}^h)^T$ and the full set of dropout weights by $\mathbf{M} = (\mathbf{m}_1, \ldots, \mathbf{m}_K)$. Thus, any neural network $\hat{f}(\mathbf{x})$ with dropout layers essentially has two sets of parameters: the full set of learnable weights $\mathbf{W}$ and the set of dropout weights $\mathbf{M}$:

$$\hat{f}(\mathbf{x}) = \hat{f}(\mathbf{x} \mid \mathbf{W}, \mathbf{M}).$$

Let us have a neural network with dropout, which was trained on some dataset giving weight estimates $\hat{\mathbf{W}}$. Then dropout weights $\mathbf{M}$ can be considered as free parameters and require selection at the time of inference

$$\hat{f}(\mathbf{x} \mid \mathbf{M}) = \hat{f}(\mathbf{x} \mid \hat{\mathbf{W}}, \mathbf{M}).$$

The originally proposed [18] and currently the standard choice is to take $\hat{\mathbf{M}} = (1 - p) \cdot \mathbf{E}$, where $\mathbf{E}$ is the matrix of all ones of the corresponding shape. Such an approach gives the fixed function $\hat{f}(\mathbf{x} \mid \hat{\mathbf{M}})$, which is known to give reasonably good performance in practice. The main intuition behind such choice is the replacement of the stochastic pre-activations S_i^h given by (2) with their expectations, which are exactly equal to (1).

Recently, it was proposed to consider dropout as a variational approximation in a specially chosen Bayesian model, see [10]. Within this approach, one can sample T i.i.d. realizations $\mathbf{M}_1, \ldots, \mathbf{M}_T \sim Bernoulli(1 - p)$ and compute approximate posterior mean and variance

$$\bar{f}_T(\mathbf{x}) = \frac{1}{T} \sum_{i=1}^{T} \hat{f}(\mathbf{x} \mid \mathbf{M}_i), \quad \bar{\sigma}_T^2(\mathbf{x}) = \frac{1}{T} \sum_{i=1}^{T} (\hat{f}(\mathbf{x} \mid \mathbf{M}_i) - \bar{f}_T(\mathbf{x}))^2.$$

The approximate posterior variance $\bar{\sigma}_T^2(\mathbf{x})$ is a natural choice for the uncertainty estimate and was successfully used in the variety of applications such as out-of-distribution detection [40] and active learning [11].

In this paper, we suggest a different approach, namely we treat $\hat{f}(\mathbf{x} \mid \mathbf{M})$ as an ensemble of models indexed by dropout masks $\mathbf{M}$. Such a view allows us to decouple inference from training and pose an intuitive question: what set of masks $\mathbf{M}_1, \ldots, \mathbf{M}_T$ should one choose in order to obtain the best uncertainty estimate $\bar{\sigma}_T^2(\mathbf{x})$?

Importantly, here we do not limit the selection of masks to be samples from standard dropout distribution, which, in principle, should allow us to obtain better estimates. However, the design of mask selection procedure is a non-trivial problem, which we discuss below in detail.

Remark 1. The standard approach in the literature is to consider an ensemble of models trained on different subsets of the data set or just from different random initializations giving the set of parameter estimates $\hat{\mathbf{W}}_1, \ldots, \hat{\mathbf{W}}_T$ and corresponding approximations $\hat{f}(\mathbf{x} \mid \hat{\mathbf{W}}_i, \hat{\mathbf{M}})$, $i = 1, \ldots, T$. Similarly, one can compute the variance $\bar{\sigma}_T^2(\mathbf{x})$, which was shown to be a reasonable uncertainty estimate in practice [5,36]. The main drawback of this approach is the need to train and store T different models, which might be very costly both in terms of computation and storage needed.

2.2 Data-Driven Mask Generation Under General Sampling Distributions

In practice, many neurons in the network are highly correlated. For example, consider a correlation matrix of neurons in a hidden layer of a fully-connected neural network, trained on the regression dataset (see Fig. 1a). The correlation matrix was computed on the test set and clearly shows groups of highly correlated neurons. Sampling masks for this layer uniformly at random might result in a high variance of pre-activations (2). As a result, the estimates for the whole network may require a significant number of samples (stochastic passes through the NN) T to converge. We illustrate this behaviour on Fig. 1b, where several hundreds of simple MC dropout estimates are required for the convergence of the log-likelihood values. It is clearly seen that a larger number of samples improves the values of log-likelihood, yet may impose computational cost too large to be used in real-world applications. However, one may expect that the knowledge about the correlations between neurons can help to sample more diverse neurons and improve the estimates.

In what follows, we consider the probabilistic generation of masks $\mathbf{m}_h$ from some distribution $P^{(h)}$ with possibly non-i.i.d. distributions of components. Similarly to the case of dropout, we suggest using an unbiased estimate of the layer-wise mean. Our main motivation is to approximately preserve the average performance of the trained network. The construction of the unbiased estimator is non-trivial and is given by celebrated Horvitz-Thompson (HT) estimator [19]:

$$S_i^h = \sum_{j=1}^{N_h - 1} \frac{1}{\pi_j^h} m_j^h w_{ij}^h O_j^{h-1}, \quad i = 1, \ldots, N_h, \tag{3}$$

where π_j^h is the marginal probability of value 1 for the random variable m_j^h.

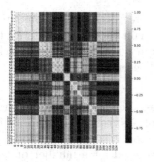

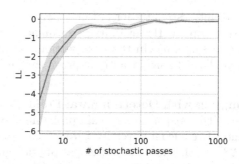

(a) Correlation matrix.

(b) Log-likelihood for MC dropout as a function of T.

Fig. 1. (a) Correlation matrix C between the outputs of the neurons in a hidden layer of the NN trained on the *naval propulsion* dataset. (b) For the same dataset log-likelihood computed via MC dropout increases with increase of the number of stochastic passes T. More than 100 samples are needed to reach convergence.

2.3   Diversity Sampling Approaches

Let us consider h-th hidden layer of the neural network with dropout. Assume that we have access to the correlations

$$C_{ij}^{(h)} = \mathrm{corr}_{\mathbf{x}}\{O_i^h(\mathbf{x}), O_j^h(\mathbf{x})\}, \; i,j = 1,\dots,N_h.$$

In practice, we compute an empirical correlation based on some set of points, which represents the data distribution well enough. As a result, we obtain the correlation matrix $C^{(h)} \in \mathbb{R}^{N_h \times N_h}$ between the neurons of the h-th hidden layer. Below we discuss several approaches to sampling neurons in a way that the correlation between sampled neurons is as small as possible. We note that instead of the correlation matrix $C^{(h)}$ one may consider the covariance matrix $K^{(h)}$ in any of the approaches described below. The properties of the methods significantly depend on the choice of the matrix, and we will perform the empirical evaluation of the methods based on each of them in the experiments.

Leverage Score Sampling. A basic approach for non-uniform sampling of rows and columns in kernel matrices is the so-called *leverage score sampling* [1]. In this approach, the neurons are sampled independently with different probabilities π_j^h:

$$\pi_j^h \sim \ell_\lambda^{(h)}(j) = \left[C^{(h)} \big(C^{(h)} + \lambda I \big)^{-1} \right]_{jj}, \; j = 1,\dots,N_h,$$

where the quantities $\ell_\lambda^{(h)}(j)$ are called *leverage scores*. This approach makes neurons from large and highly correlated clusters to be sampled less frequently. In Sect. 3, we show that leverage score sampling indeed allows obtaining better uncertainty estimates for out-of-distribution data in regression tasks compared to MC dropout. However, its performance for in-domain data is even inferior to uniform sampling. In the next section, we propose a more complex approach, which allows to significantly improve the quality of uncertainty estimation.

Sampling with Determinantal Point Processes. Determinantal Point Processes (DPPs) [24] are specific probability distributions over configurations of points that encode diversity through a kernel function. They were introduced in [28] for the needs of statistical physics and were used for a number of ML applications, see [24] for an overview. DPP can be seen as a probabilistic MaxVol algorithm [14] of finding a maximal-volume submatrix.

We use correlation matrix $C^{(h)}$ as the likelihood kernel for DPP. Then, given a set S of selected points for a mask distribution $\mathbf{m}_h \sim DPP\big(C^{(h)}\big)$, we obtain

$$\mathbb{P}[\mathbf{m}_h = S] = \frac{\det\big[C_S^{(h)}\big]}{\det\big[C^{(h)} + I\big]}, \quad h = 1, \ldots, K,$$

where $C_S^{(h)} = \big[C_{ij}^{(h)}, \ i, j \in S\big]$, i.e., a square submatrix of $C^{(h)}$ obtained by keeping only rows and columns indexed by S.

To better understand the DPP, let us come back to the correlation matrix depicted in Fig. 1a. The probability for DPP to take highly correlated neurons into the sample S is low as, in this case, the corresponding determinant $\det C_S^{(h)}$ will have a small value. Thus, DPP tends to sample neurons from different clusters, increasing an overall diversity.

From computational point of view, DPP-sampling requires $O(N_h^3)$ operations for generating each sample. It is quite expensive but completely viable even for modern large networks which usually have up to 1024 neurons in fully-connected layers. Importantly, masks can be precomputed once, and then the same masks are used on the inference stage for every test sample with no additional overhead. Also, computations in last fully-connected layers with dropout usually require only few percents of the total computational budget in ImageNet-size networks. Therefore, a computational overhead caused by the DPP-sampling does not have a significant impact on the inference time.

K-DPP. The k-DPP [24] is a variation of the DPP, conditioned to produce samples of fixed size $|S| = k$. With the cost of introducing an additional parameter, it allows us to tune the sampling procedure as the choice of k apparently has a significant influence on the result. In this work, we use for the h-th layer $k^{(h)} = (1 - p)N_h$, so that the number of neurons in the sample is equal to the mean number of neurons in the sample of MC-Dropout. In the case of k-DPP, the computation of the marginal probabilities π_j^h for HT-estimator (3) is non-trivial and requires the separate optimization procedure, see the details in [2].

2.4   Diversification for Uncertainty Estimation in Classification

For regression, the variance of prediction is a standard uncertainty measure. However, uncertainty estimation for classification is, in some sense, more challenging than for regression as there is no obvious candidate for uncertainty measure.

Let us define the average probability for the class prediction by ensemble members $\bar{p}_T(y = c \mid \mathbf{x}) = \frac{1}{T} \sum_{i=1}^{T} p(y = c \mid \mathbf{x}, \mathbf{M}_i)$. The standard uncertainty measure usually considered in the literature is

$$s(\mathbf{x}) = 1 - \max_c \; \bar{p}_T(y = c \mid \mathbf{x}),$$

which is based solely on the mean probabilities predicted by the ensemble. While providing good results in practice [3,37] it doesn't use the information about the variation of predictions between ensemble members.

In our work, we consider *BALD* [20] uncertainty measure and combine it with different sampling schemes considered above. BALD is equal to the mutual information between outputs and model parameters:

$$I(\mathbf{x}) = H(\mathbf{x}) - \frac{1}{T} \sum_{c=1}^{C} \sum_{i=1}^{T} -p(y = c \mid \mathbf{x}, \mathbf{M}_i) \log\big(p(y = c \mid \mathbf{x}, \mathbf{M}_i)\big),$$

where $H(\mathbf{x}) = - \sum_{c=1}^{C} \bar{p}_T(y = c \mid \mathbf{x}) \log(\bar{p}_T(y = c \mid \mathbf{x}))$ is an entropy of the ensemble mean. Importantly, BALD values are directly linked with the diversity of the ensemble members, and therefore are well suited for combination with our approach.

3   Experiments

3.1   Uncertainty Estimation for Regression

Models and Metrics. For the experiments, we consider MC dropout as a baseline and all the proposed UE methods discussed in the Sect. 2.3: leverage score sampling, DPP and k-DPP. We present the results for leverage score sampling and DPP based on correlation matrix and k-DPP based on covariance matrix as such a choices give consistently better results compared to an alternative. For leverage score sampling we deliberately choose $\lambda = 1$ to make it working with de-facto the same matrix as DPP-based methods. All the regression models were trained with RMSE as a loss function. We used feed-forward NNs with 3 hidden layers (128-128-64 neurons) and leaky ReLU activation function [27]. For DPP-based methods, we use the DPPy implementation provided in [13].

We should note that we do not compare with fully Bayesian approaches as we are focusing on the solutions applicable to the standard dropout-based models without changing model architecture and training procedure. Following [17,22], we use log-likelihood of Gaussian distribution with mean and variance computed by different methods as a quality measure.

On top of single models, we also consider a straightforward ensemble approach with NNs trained exactly the same way as single models but from different

Table 1. Summary of the UCI datasets used in experiments, see [7].

Dataset name	Naval propulsion	Concrete	Boston housing	Kin8nm	Ccpp	Red wine
Samples	11934	1030	506	8192	9568	1599
Features	16	8	13	8	4	11

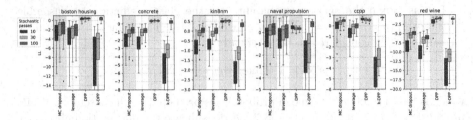

Fig. 2. Log-likelihood metric across various UCI datasets for NN UE models with different number of stochastic passes $T = 10, 30, 100$. DPP and k-DPP give better results compared to other methods with DPP working well already for $T = 10$ and consistently showing lower variance.

random initializations. Our experiments show that uncertainty estimates based on ensembling of networks without sampling in individual networks doesn't work for well for the considered regression datasets.

Experiments on Regression Datasets. Similarly to [22], we run a series of experiments on various regression datasets, see Table 1 for the full list of datasets. We start with in-domain uncertainty estimation: for each dataset, random 50% of points were used for training and other 50% for testing. The log-likelihood values are averaged over testing set. Multiple experiments are done via 5 random train-test splits, 2-fold cross-validation and 5 runs of the training procedures for every model (resulting in 50 average log-likelihood values contributing to each boxplot). Uncertainty estimates were computed for different number of stochastic passes $T = 10, 30$ and 100 for every model.

We show the resulting distributions of log-likelihood values for each dataset on Fig. 2. We observe that either DPP or k-DPP always show the best results. Most importantly, DPP works very well already for small number of stochastic passes $T = 10$ and consistently has low variance which is extremely important for practical usage.

We also performed an experiment with out-of-distribution (OOD) data. To generate OOD data we pick a random feature and split the data into the train set and OOD set by the median value on this feature. The experiments were run for 5 different splits. For OOD data good uncertainty estimates should have on average higher values compared to in-domain data. Table 2 provides for *concrete* dataset the percentages of OOD points with UE values higher than α percentile of UE distribution for training data ($\alpha = 80\%, 90\%, 95\%$). The resulting numbers

Table 2. Percentages of OOD points with UE values higher than specified percentile of UE distribution for training data for *concrete* dataset. DPP and k-DPP show the best results based on average values (top-2 average values are put in **bold**). For all the methods $T = 100$.

Percentile	MC dropout	Leverage	DPP	k-DPP
80	55.0 ± 27.6	61.3 ± 27.7	$\mathbf{70.4 \pm 26.0}$	$\mathbf{71.9 \pm 28.0}$
90	46.0 ± 30.7	52.9 ± 30.8	$\mathbf{59.6 \pm 30.1}$	$\mathbf{60.8 \pm 33.7}$
95	40.6 ± 32.1	46.5 ± 33.1	$\mathbf{52.1 \pm 32.9}$	$\mathbf{51.8 \pm 36.3}$

should be considered with a significant grain of salt due to their high variance but still DPP and k-DPP show the best results based on average values.

3.2 Uncertainty Estimation for Classification

Data, Models and Metrics. In this section, we aim to show the applicability of the proposed methods to the classification tasks. We take *BALD* [20] as an uncertainty estimate. We consider three datasets: MNIST, which is a toy dataset of handwritten digits [26], CIFAR-10, which is a 10-class image dataset with simple objects [23], and ImageNet [6], the large scale image classification dataset. Importantly, for MNIST we use only 500 train samples, otherwise the models would have too good accuracy and uncertainty estimation for in-domain data would not be relevant. For CIFAR-10 we use 50'000 samples for training and 10'000 for testing. For the MNIST dataset, we use a simple convolutional neural network with two convolutional layers, max-pooling and two fully connected layers. For the CIFAR-10 we use a more powerful network with 6 convolutional layers and batch normalization. Finally, for ImageNet we use the pre-trained ResNet-18 neural network [16] from PyTorch [33]. Dropout with rate $p = 0.5$ is used before the last fully-connected layer in all the cases. $T = 100$ stochastic passes were made for every model. The experiments are repeated three times with different seeds for the models.

Experimental Results. For in-domain uncertainty estimation the results are presented via UE-accuracy curve, see Fig. 3. It assumes that samples with lower uncertainty will be classified with a higher average accuracy. It can be clearly seen that DPP significantly outperforms all the competitors on every dataset. We should emphasize that the superiority of DPP is especially strong for ImageNet, where the usage of DPP required only 2% computational overhead compared to MC dropout according to our experiments.

We also consider detection of out-of-distribution samples which is one of the important problems for the uncertainty estimation. As OOD samples we use fashion-MNIST [41] and SVHN images [32] for MNIST and CIFAR-10 correspondingly. We use count-vs-uncertainty curve and expect there should be few

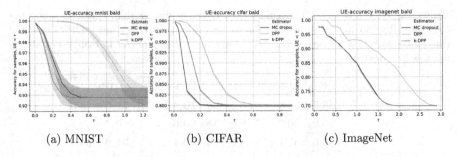

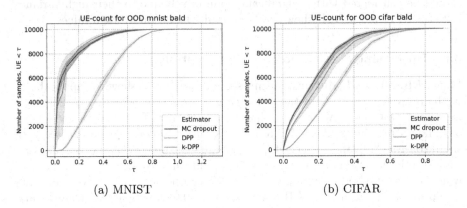

Fig. 3. UE-accuracy curve (the higher curve – the better). We select the samples with low uncertainty to assure that the accuracy is higher for them.

Fig. 4. Count-vs-uncertainty curve for out-of-distribution data (the lower curve – the better).

points with the low uncertainty for good uncertainty estimation methods. The results are presented in Fig. 4. We see that DPP-based approach allows to detect the OOD samples better for the both considered datasets.

4   Related Work

Dropout [18,38] has emerged in recent years as a technique to prevent the overfitting in deep and overparametrized neural networks. Over the years, it obtained theoretical explanations as an averaged ensembling technique [38], a Bernoulli realization of the corresponding Bayesian neural network [10] and a latent variable model [30]. It was shown in [9,31] that using dropout at the prediction stage (i.e., stochastic forward passes of the test samples through the network, also referred to as *MC dropout*) leads to unbiased Monte-Carlo estimates of the mean and variance for the corresponding Bayesian neural network trained using variational inference. These uncertainty estimates were shown to be efficient in different scenarios [9,39].

Training an ensemble of models and uncertainty estimation by their disagreement is another common approach [25]. It is shown that with few models in an ensemble, you can get robust and useful calibrated results [5], outperforming MC dropout in active learning and error detection. The main disadvantage of ensembles is the necessity to train multiple model instances. However, it was addressed in recent works [12,21,29] which consider different strategies for speeding up ensemble construction. Recently, it was shown that improving diversity of ensemble members improves the quality of the resulting uncertainty estimates [22]. We also mention recent works which thoroughly investigate in-domain [3] and out-of-domain [37] uncertainty estimation in classification for the case of maximum probability uncertainty estimate.

5   Conclusions

We have proposed a new approach that strengthens the dropout-based uncertainty estimation for neural networks. Instead of randomly sampling the dropout masks on the inference stage, we sample special sets of diverse neurons via determinantal point processes that utilize the information about the correlations of neurons in the inner layers. Numerical experiments on a wide range of regression and classification tasks show that uncertainty estimates based our approach outperform the MC dropout and other baselines with a significant margin. A combination of dropout-based inference with ensembling of several models allows to further improve the quality of the proposed uncertainty estimates and achieve state-of-the-art performance. From the practical perspective, our method is simple to implement as it does not require any modifications to the neural network architecture and the training process. Importantly, the proposed uncertainty estimates have high quality even for a small number of stochastic passes through the network making the inference stage even faster in practice.

We expect that the proposed methods of dropout mask sampling may also be used on the training stage, leading to more robust and efficient models. Another compelling direction of further research is approximate DPP sampling, which may increase the sampling speed of the proposed approaches, making them more production-friendly, as in [35].

The code reproducing the experiments is available at Github[1].

Acknowledgements. The research was carried out at Skoltech and supported by the Russian Science Foundation (project no. 21-11-00373). The authors want to thank Nikita Mokrov for useful discussions. M.P. and K.F. acknowledge the use of "Zhores" supercomputer [42] for obtaining the part of results presented in this paper.

References

1. Alaoui, A., Mahoney, M.W.: Fast randomized kernel ridge regression with statistical guarantees. In: NIPS, pp. 775–783 (2015)

---

[1] https://github.com/stat-ml/dpp-dropout-uncertainty/.

2. Amblard, P.O., Barthelmé, S., Tremblay, N.: Subsampling with k determinantal point processes for estimating statistics in large data sets. In: 2018 IEEE Statistical Signal Processing Workshop (SSP), pp. 313–317. IEEE (2018)
3. Ashukha, A., Lyzhov, A., Molchanov, D., Vetrov, D.: Pitfalls of in-domain uncertainty estimation and ensembling in deep learning. In: ICLR (2019)
4. Begoli, E., Bhattacharya, T., Kusnezov, D.: The need for uncertainty quantification in machine-assisted medical decision making. Nat. Mach. Intell. 1(1), 20–23 (2019)
5. Beluch, W.H., Genewein, T., Nürnberger, A., Köhler, J.M.: The power of ensembles for active learning in image classification. In: CVPR, pp. 9368–9377 (2018)
6. Deng, J., Dong, W., Socher, R., Li, L.J., Li, K., Fei-Fei, L.: ImageNet: a large-scale hierarchical image database. In: CVPR, pp. 248–255. IEEE (2009)
7. Dua, D., Taniskidou, E.K.: UCI machine learning repository (2017). http://archive.ics.uci.edu/ml
8. Feng, D., Rosenbaum, L., Dietmayer, K.: Towards safe autonomous driving: capture uncertainty in the deep neural network for lidar 3D vehicle detection. In: ITSC, pp. 3266–3273 (2018)
9. Gal, Y.: Uncertainty in deep learning. University of Cambridge (2016)
10. Gal, Y., Ghahramani, Z.: Dropout as a Bayesian approximation: representing model uncertainty in deep learning. In: ICML, pp. 1050–1059 (2016)
11. Gal, Y., Islam, R., Ghahramani, Z.: Deep Bayesian active learning with image data. In: ICML, pp. 1183–1192 (2017)
12. Garipov, T., Izmailov, P., Podoprikhin, D., Vetrov, D.P., Wilson, A.G.: Loss surfaces, mode connectivity, and fast ensembling of DNNs. In: NeurIPS, pp. 8789–8798 (2018)
13. Gautier, G., Polito, G., Bardenet, R., Valko, M.: DPPy: DPP sampling with Python. JMLR 20(180), 1–7 (2019)
14. Goreinov, S., Oseledets, I., Savostyanov, D., Tyrtyshnikov, E., Zamarashkin, N.: How to find a good submatrix. In: Matrix Methods: Theory, Algorithms and Applications: Dedicated to the Memory of Gene Golub, pp. 247–256 (2010)
15. Guo, C., Pleiss, G., Sun, Y., Weinberger, K.Q.: On calibration of modern neural networks. In: Proceedings ICML, pp. 1321–1330 (2017)
16. He, K., Zhang, X., Ren, S., Sun, J.: Deep residual learning for image recognition. In: CVPR, pp. 770–778 (2016)
17. Hernández-Lobato, J.M., Adams, R.: Probabilistic backpropagation for scalable learning of Bayesian neural networks. In: ICML, pp. 1861–1869 (2015)
18. Hinton, G.E., Srivastava, N., Krizhevsky, A., Sutskever, I., Salakhutdinov, R.R.: Improving neural networks by preventing co-adaptation of feature detectors. arXiv arXiv:1207.0580 (2012)
19. Horvitz, D.G., Thompson, D.J.: A generalization of sampling without replacement from a finite universe. JASA 47(260), 663–685 (1952)
20. Houlsby, N., Huszár, F., Ghahramani, Z., Lengyel, M.: Bayesian active learning for classification and preference learning. arXiv preprint arXiv:1112.5745 (2011)
21. Izmailov, P., Maddox, W., Kirichenko, P., Garipov, T., Vetrov, D., Wilson, A.G.: Subspace inference for Bayesian deep learning (2019)
22. Jain, S., Liu, G., Mueller, J., Gifford, D.: Maximizing overall diversity for improved uncertainty estimates in deep ensembles. In: Proceedings of the AAAI Conference on Artificial Intelligence, vol. 34, pp. 4264–4271 (2020)
23. Krizhevsky, A., Hinton, G., et al.: Learning multiple layers of features from tiny images. Technical report TR-2009, University of Toronto (2009)
24. Kulesza, A., Taskar, B., et al.: Determinantal point processes for machine learning. Found. Trends® Mach. Learn. 5(2–3), 123–286 (2012)

25. Lakshminarayanan, B., Pritzel, A., Blundell, C.: Simple and scalable predictive uncertainty estimation using deep ensembles. In: NIPS, pp. 6402–6413 (2017)
26. LeCun, Y.: The MNIST database of handwritten digits (1998). https://yannlecun.com/exdb/mnist/
27. Maas, A.L., Hannun, A.Y., Ng, A.Y.: Rectifier nonlinearities improve neural network acoustic models. In: Proceedings of the ICML 2013, vol. 30, p. 3 (2013)
28. Macchi, O.: The coincidence approach to stochastic point processes. Adv. Appl. Probab. **7**(1), 83–122 (1975)
29. Maddox, W.J., Izmailov, P., Garipov, T., Vetrov, D.P., Wilson, A.G.: A simple baseline for Bayesian uncertainty in deep learning. In: NeurIPS, pp. 13132–13143 (2019)
30. Maeda, S.: A Bayesian encourages dropout. arXiv arXiv:1412.7003 (2014)
31. Nalisnick, E., Hernandez-Lobato, J.M., Smyth, P.: Dropout as a structured shrinkage prior. In: International Conference on Machine Learning, pp. 4712–4722 (2019)
32. Netzer, Y., Wang, T., Coates, A., Bissacco, A., Wu, B., Ng, A.Y.: Reading digits in natural images with unsupervised feature learning (2011)
33. Paszke, A., et al.: PyTorch: an imperative style, high-performance deep learning library. In: NeurIPS, pp. 8024–8035 (2019)
34. Settles, B.: Active Learning. Synthesis Lectures on Artificial Intelligence and Machine Learning, vol. 6, no. 1, pp. 1–114 (2012)
35. Shelmanov, A., Tsymbalov, E., Puzyrev, D., Fedyanin, K., Panchenko, A., Panov, M.: How certain is your transformer? In: Proceedings of the 16th Conference of the European Chapter of the Association for Computational Linguistics: Main Volume, pp. 1833–1840 (2021). https://doi.org/10.18653/v1/2021.eacl-main.157
36. Smith, J.S., Nebgen, B., Lubbers, N., Isayev, O., Roitberg, A.E.: Less is more: sampling chemical space with active learning. J. Chem. Phys. **148**(24), 241733 (2018)
37. Snoek, J., et al.: Can you trust your model's uncertainty? evaluating predictive uncertainty under dataset shift. In: NeurIPS, pp. 13969–13980 (2019)
38. Srivastava, N., Hinton, G., Krizhevsky, A., Sutskever, I., Salakhutdinov, R.: Dropout: a simple way to prevent neural networks from overfitting. J. Mach. Learn. Res. **15**(1), 1929–1958 (2014)
39. Tsymbalov, E., Panov, M., Shapeev, A.: Dropout-based active learning for regression. In: International Conference on Analysis of Images, Social Networks and Texts, pp. 247–258 (2018)
40. Vyas, A., Jammalamadaka, N., Zhu, X., Das, D., Kaul, B., Willke, T.L.: Out-of-distribution detection using an ensemble of self supervised leave-out classifiers. In: Ferrari, V., Hebert, M., Sminchisescu, C., Weiss, Y. (eds.) ECCV 2018. LNCS, vol. 11212, pp. 560–574. Springer, Cham (2018). https://doi.org/10.1007/978-3-030-01237-3_34
41. Xiao, H., Rasul, K., Vollgraf, R.: Fashion-MNIST: a novel image dataset for benchmarking machine learning algorithms (2017)
42. Zacharov, I., et al.: "Zhores"—petaflops supercomputer for data-driven modeling, machine learning and artificial intelligence installed in Skolkovo Institute of Science and Technology. Open Eng. **9**(1), 512–520 (2019)

Learning to Generate Synthetic Training Data Using Gradient Matching and Implicit Differentiation

Dmitry Medvedev[✉] and Alexander D'yakonov

Lomonosov Moscow State University, Moscow, Russia
dm.medvedev97@gmail.com

Abstract. Using huge training datasets can be costly and inconvenient. This article explores various data distillation techniques that can reduce the amount of data required to successfully train deep networks. Inspired by recent ideas, we suggest new data distillation techniques based on generative teaching networks, gradient matching, and the Implicit Function Theorem. Experiments with the MNIST image classification problem show that the new methods are computationally more efficient than previous ones and allow to increase the performance of models trained on distilled data.

Keywords: Data distillation · Gradient matching · Implicit differentiation · Generative teaching network

1 Introduction

In machine learning, the purpose of data distillation [1] is to compress the original dataset while maintaining the performance of the models trained on it. The generalizability of the dataset is also needed. By this we mean the ability to train models of architectures that were not involved in the distillation process. Since training with less data is usually faster, distillation can be useful in practice. For example, it can be used to speed up a neural architecture search (NAS) task. Acceleration is achieved through the faster training of candidates. In many recent works [1,3,5–7], distillation is formulated as an optimization problem with the objects of a new dataset as parameters for optimization. Therefore, to distill the dataset for an image classification task, pixels of images have to be optimized. First, all new objects are initialized with random noise, then these objects are used to train a student (i.e., a randomly selected network). Then the student misclassification loss is calculated on real data. Finally, a gradient descent step is used to update the synthetic objects. Gradients can be calculated by backpropagating the error through the entire student's learning process. The step of this procedure can be very time-consuming and memory-intensive, so there is a need for an alternative. In [2], the authors use the Implicit Function

© The Author(s), under exclusive license to Springer Nature Switzerland AG 2022
E. Burnaev et al. (Eds.): AIST 2021, CCIS 1573, pp. 138–150, 2022.
https://doi.org/10.1007/978-3-031-15168-2_12

Theorem to solve the memory consumption problem. In [3], the data distillation problem has been reformulated to use gradient matching loss and speed up the optimization of synthetic objects and reduce memory usage. There is an alternative to optimizing the pixels of synthetic data. In [4], the authors suggest to optimize parameters of the generator model (a generative teaching network or GTN) to produce synthetic data from noise and labels. The disadvantage is that the authors used backpropagation through the learning process for optimization. Inspired by recent ideas in the field of data distillation, we propose replacing it with gradient matching or with implicit differentiation to make the procedure less computationally expensive. We have found that this allows not only to reduce memory costs but also to create more efficient and generalizable datasets. In addition, we investigate the use of augmentation in the distillation procedure and in models' learning on distilled data.

The paper is divided into 7 sections. We first analyse the first data distillation algorithm [1] and discuss its problems in Sect. 2. A brief description of the algorithms for implicit differentiation [2] and gradient matching [3] can be found in Sects. 3 and 4. Section 5 presents the generative teaching network architecture that we use in our work. Section 6 contains the results of experiments with the MNIST image classification benchmark. In Sect. 6.1 we compare the results of all the described distillation methods, limiting the distillation time to a constant. In Sects. 6.2 and 6.3 we show results of new distillation techniques when training a generator with gradient matching and implicit differentiation, respectively. In Sect. 6.4 we study the use of augmentation by distillation, and in Sect. 6.5 we check the generalization of the data obtained with the new methods. Finally, we present our findings in Sect. 7. The code can be found on our GitHub page.[1]

2   Backpropagation Through the Student's Learning Process

Let λ be teacher parameters. These can be either GTN network's parameters, or synthetic objects' parameters (e.g. pixels of synthetic images). To update λ, we must first train the student network θ on synthetic data, minimizing the task specific loss $\mathcal{L}_S$ (e.g. cross-entropy), and then get the loss on real data $\mathcal{L}_T$. To take care of generalizability, student's initialization goes from preset distribution $p(\theta_0)$. Afterall, the optimization problem for λ can be formulated as follows:

$$\lambda^* := \underset{\lambda}{\operatorname{argmin}} \ \mathbb{E}_{\theta_0 \sim p(\theta_0)} \mathcal{L}_T^*, \text{ where} \tag{1}$$

$$\mathcal{L}_T^* := \mathcal{L}_T(\theta^*(\lambda)), \qquad \theta^*(\lambda) := \underset{\theta}{\operatorname{argmin}} \ \mathcal{L}_S(\lambda, \theta).$$

To resolve the first problem (1) we can calculate gradient of $\mathcal{L}_T$ with respect to λ to do the gradient descent step. In this work, we use cross-entropy loss as

---

[1] https://github.com/dm-medvedev/EfficientDistillation.

$\mathcal{L}_\mathcal{T}$ and there is an explicit dependence only on θ and parameters of real data, so $\frac{\partial \mathcal{L}_\mathcal{T}}{\partial \lambda} = 0$ and $\frac{\partial \mathcal{L}_\mathcal{T}^*}{\partial \lambda} = \frac{\partial \mathcal{L}_\mathcal{T}}{\partial \theta} \frac{\partial \theta^*}{\partial \lambda}$. Thus, the main part is the calculation of $\frac{\partial \theta^*}{\partial \lambda}$. Where the dependence of θ^* on λ comes from a student's training procedure. The first distillation algorithm was suggested in [1] and it is based on the assumption that the student's learning procedure is differentiable. This means that we can backpropagate gradient through it. We will denote it as **unroll**. This algorithm can be implemented using the Higher library [10]. It allows to backpropagate through many optimizers, in our paper we use SGD with momentum [8]. This distillation method is both time and space consuming. To perform a single step of updating λ it is necessary to perform N student optimization steps, while all intermediate results (copies of the student weights) must be stored in memory. There is also a problem with the generalization of resulting synthetic dataset, the performance of models whose architectures were not involved in the distillation process is much lower. This negative effect can be mitigated by sampling the initialization and student architecture.

Note that the procedure of student's training on the resulting synthetic dataset can be carried out in different ways. New data, parameterized with λ, can be used as a single large batch or it can be split into several smaller ones. This split can be useful to reduce memory consumption per training step. Instead of random sampling of distilled objects, the authors of the original work propose to attach each of them to a specific batch. These batches would have the same order in each epoch. In our paper, we use the same schemes. Let ic (input count) be the number of batches of the synthetic dataset, note that it must be divisor of N. In our experiments we try limit values $ic = 1$ and $ic = 10$.

3   Implicit Differentiation

This method suggested in [2] is based on the Implicit Function Theorem:

Theorem 1 (Cauchy, Implicit Function Theorem). *Let* $\frac{\partial \mathcal{L}_\mathcal{S}}{\partial \theta}(\lambda, \theta) : \Lambda \times \Theta \to \Theta$, *be a continuously differentiable function. Fix a point* (λ', θ') *with* $\frac{\partial \mathcal{L}_\mathcal{S}}{\partial \theta}(\lambda', \theta') = 0$. *If the Jacobian matrix* $\frac{\partial^2 \mathcal{L}_\mathcal{S}}{\partial \theta^2}$ *is invertible, then there exists an open set* $U \subseteq \Lambda$ *containing* λ' *such that there exists a unique continuously differentiable function* $\theta^* : U \to \Theta$, *such that* $\theta^*(\lambda') = \theta'$ *and* $\forall \lambda \in U$, $\frac{\partial \mathcal{L}_\mathcal{S}}{\partial \theta}(\lambda, \theta^*(\lambda)) = 0$. *Moreover, the partial derivatives of* θ^* *in* U *are given by the matrix product:*

$$\frac{\partial \theta^*}{\partial \lambda}(\lambda) = -\left[\frac{\partial^2 \mathcal{L}_\mathcal{S}}{\partial \theta^2}(\lambda, \theta^*(\lambda))\right]^{-1} \frac{\partial^2 \mathcal{L}_\mathcal{S}}{\partial \theta \partial \lambda}(\lambda, \theta^*(\lambda)). \tag{2}$$

So, if there was an efficient way to invert the matrix, we would simply have used (2), after the student θ has reached a local minimum, assuming $\frac{\partial \mathcal{L}_\mathcal{S}}{\partial \theta}(\lambda, \theta^*(\lambda)) \approx 0$. But the inversion operation is time costly, so the authors used the approximation by the Neumann series taking the first few elements and controlling convergence with a hyperparameter α (see (3)).

The resulting algorithm (see Algorithm 1) has no problems with memory consumption since there is no need to store copies of the student θ. And, despite the several subsequent approximations, the experimental results show that the method has a competitive performance (see Table 4). Note that **grad** in Algorithm 1 denotes the dot product between the Jacobian of the given function (**func**) at the given point (**wrt**) and a vector (**vec**). Another interesting detail of this method is that there is no dependence on which optimizer is used to train the student, and on the order (curriculum) of batches of synthetic data. So, in our paper we only use a single large batch of synthetic data. The original work [2] lacks a detailed description of the experimental results, so it can be found in our paper (see Sect. 6.3). We used the open-source code[2] as the basis for the implementing the method.

$$
\left[\frac{\partial^2 \mathcal{L}_S}{\partial \theta^2} (\lambda, \theta^*(\lambda)) \right]^{-1} \approx \alpha \sum_{j=0}^{N} \left[I - \alpha \frac{\partial^2 \mathcal{L}_S}{\partial \theta^2} (\lambda, \theta^*(\lambda)) \right]^j . \tag{3}
$$

Algorithm 1. Distillation with implicit differentiation.

1: **Input:** teacher's parameters λ, student's initialization distribution $p(\theta_0)$, the number of distillation epochs K, the number of student's learning steps ζ_θ, real data $\mathcal{T}$, learning rate η.
2: **for** $k = 1, ..., K$ **do**
3: $\quad \mathcal{B}^\mathcal{T} \sim \mathcal{T}, \quad \theta \sim p(\theta_0)$
4: $\quad$ **for** $n = 1, ..., \zeta_\theta$ **do**
5: $\quad\quad \theta \mathrel{-}= \eta \frac{\partial \mathcal{L}_S(\lambda, \theta)}{\partial \theta}$
6: $\quad \mathcal{L}_\mathcal{T} = ClassificationLoss(\mathcal{B}^\mathcal{T}, \theta)$
7: $\quad v = \frac{\partial \mathcal{L}_\mathcal{T}}{\partial \theta}; \quad p = v$
8: $\quad$ **for** $j = 1, ..., N$ **do** $\qquad\qquad\qquad$ ▷ N — number of elements in (3)
9: $\quad\quad v \mathrel{-}= \alpha \cdot \mathbf{grad}\left(\mathbf{func} = \frac{\partial \mathcal{L}_S}{\partial \theta}, \mathbf{wrt} = \theta, \mathbf{vec} = v\right)$
10: $\quad\quad p \mathrel{+}= v$
11: $\quad \nabla_\lambda \mathcal{L}_\mathcal{T} = -\alpha \cdot \mathbf{grad}\left(\mathbf{func} = \frac{\partial \mathcal{L}_S}{\partial \theta}, \mathbf{wrt} = \lambda, \mathbf{vec} = p\right)$
12: $\quad$ **Update**$(\lambda, \nabla_\lambda \mathcal{L}_\mathcal{T})$ $\qquad\qquad\qquad$ ▷ update with any optimizer
$\quad$ **return** λ

---

4 Gradient Matching

The gradient matching method (**GM**) was proposed in [3], and it solves a different problem than the general one (1). The main difference is that we want not only to train the student θ to achieve a good performance on real data but also to get such a solution as if it was trained on real data. To formulate this let $D(\nabla_\theta \mathcal{L}_S, \nabla_\theta \mathcal{L}_\mathcal{T})$ be the function of how close one tensor is to another.

---

[2] https://github.com/AvivNavon/AuxiLearn.

The distance function D is just the sum (in our paper for GTN experiments we used the mean) of the cosine distance functions for each student layer θ^l. Let A and B be gradient tensors with respect to layer parameters. Let i be the index of the output axis (e.g. for a convolutional layer this is the index of the output channel). A_i and B_i are flat gradient vectors corresponding to each output element indexed by i. The most interesting detail here is that the authors [3] suggest to update λ after each step of student optimization, so now we don't need to wait until it reaches a local minimum, as it was before. The authors also propose not to store student copies and to minimize $D(\nabla_\theta \mathcal{L}_S(\lambda, \theta_{t-1}), \nabla_\theta \mathcal{L}_T(\theta_{t-1}))$ for each step separately. So there is no backpropagation through opt_θ. Both of these proposals make the gradient matching method very computational effective.

$$\lambda^* = \underset{\lambda}{\arg\min} \, \mathbb{E}_{\theta_0 \sim P_{\theta_0}} \left[\sum_{n=1}^{N-1} D\big(\nabla_\theta \mathcal{L}_S(\lambda, \theta_n), \nabla_\theta \mathcal{L}_T(\theta_n)\big) \right], \quad \text{where:} \tag{4}$$

$$D(\nabla_\theta \mathcal{L}_S, \nabla_\theta \mathcal{L}_T) = \sum_{l=1}^{L} d(\nabla_{\theta^l} \mathcal{L}_S, \nabla_{\theta^l} \mathcal{L}_T), \quad d(A, B) = \sum_{i=1}^{\dim(A)} \left(1 - \frac{A_i \cdot B_i}{\|A_i\|\|B_i\|} \right)$$

Algorithm 2. Gradient matching.

1: **Input**: teacher's parameters λ and synthetic objects $\mathcal{S}(\lambda)$, student's initialization distribution $p(\theta_0)$, the number of distillation epochs K, the number of student's learning steps ζ_θ, real data $\mathcal{T}$, learning rate η_θ, the number of inner loop steps N.
2: **for** $k = 0, ..., K - 1$ **do**
3: $\quad \theta_0 \sim p_{\theta_0}$
4: $\quad$ **for** $n = 0, ..., N - 1$ **do**
5: $\quad\quad \mathcal{B}^T \sim \mathcal{T}, \quad \mathcal{B}^S \sim \mathcal{S}(\lambda)$
6: $\quad\quad \mathcal{L}_T = ClassificationLoss(\mathcal{B}^T, \theta_n), \quad \mathcal{L}_S = ClassificationLoss(\mathcal{B}^S, \theta_n)$
7: $\quad\quad \mathcal{L}(\lambda) = D(\nabla_\theta \mathcal{L}_S(\lambda, \theta_n), \nabla_\theta \mathcal{L}_T(\theta_n))$
8: $\quad\quad$ **Update**$(\lambda, \nabla_\lambda \mathcal{L}(\lambda))$
9: $\quad\quad \theta_{n+1} \leftarrow opt_\theta(\mathcal{L}_S(\lambda, \theta_n), \zeta_\theta, \eta_\theta)$
10: **Output**: λ

The peculiarity of this loss function is that the gradient of one synthetic object depends on other objects from the same batch, because of a normalization operation in the d equation (4). It makes the optimization problem harder and can cause negative effects (see Table 2). So authors decided to distill objects separately for each class. Note that the gradient matching is independent of the student training optimization algorithm. There is only one assumption that the direction should be based on the gradient. Another aspect is that the curriculum (the order of the synthetic batches in the student's learning procedure) can be learned with this distillation method. We used an open-source code[3] as the implementation of this method.

5 Generative Teaching Network

The idea first appeared in [4], where the authors suggested to use the generator as the teacher λ. The input of the generator is a concatenation of noise and one hot encoded label (for conditional generation). In the original paper, the authors use backpropagation through the student's learning process to train the generator, which is inconvenient for practical use due to high memory consumption, so in our paper, we show that the same or even better results can be achieved more efficiently by using gradient matching or implicit differentiation. Experimental results in [4] show that using a generator can help to improve students' performance. In our paper, we check if we can improve distillation performance using larger generators. Note that the size of the generator in our experiments is controlled by the k hyperparameter (see Fig. 1). The generator consists of two linear layers and two convolutional layers. The output size of the first layer is k. And $\lfloor k/2 \rfloor \times$ width $\times$ height of picture is the output size of the second layer. $\lfloor k/4 \rfloor$ is the number of output channels of the first convolution. Hereinafter, unless otherwise indicated, we use the following notation: **DD** (Data Distillation) is a distillation, when the parameters of the teacher λ are pixels of synthetic images, and **GTN** is a distillation using a generator. Note that the generator has two modes: **GTN-rnd** is a generator with random noise as input, (**GTN-lrn**) is a generator with a learned input.

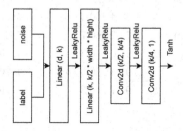

Fig. 1. Generator's architecture; k is a hyperparameter to control network's size, $d = 64$ is a generator's input.

6 Experiments

6.1 Distillation with Time Limit

The neural architecture search (NAS) is one of the most promising areas for distillation and it is important to note that the time spent on distillation should

---

[3] https://github.com/VICO-UoE/DatasetCondensation.

be added to the time spent on the NAS, this idea was also mentioned in the review[4] of [4]. So, in this section, we check the performance of all known distillation methods. We think that it is fair to distill the data by all methods for the same limited time. We have chosen a time limit of ≈15 min, and it is based on common sense and the time spent on the NAS in similar experiments [3]. Note that this limit may not be accurate, as the distillation takes an integer number of steps, where each step takes a non-deterministic time. To check the performance we use the following scheme. First we train teacher λ with three restarts. The number of steps is determined by the time limit indicated above. Then, to get the final results we train five randomly initialized students θ for each of the three teachers. Each student's training takes 1000 optimization steps. In our work we use the MNIST [9] benchmark and make the same preparations as in [4]. We extract part of the training data for validation (10 thousand images) and use it to get the best teacher hyperparameters. We use $|\mathcal{B}^T| = 256$ batch size of training data. For the most of our experiments we use ConvNet [12] as a student. As student's optimizer we use SGD with momentum with the same parameters as suggested in [3]. We use the same teacher optimizers as in the original papers [1,3,4]. The volume of synthetic data can be controlled by the *ipc* (images per class) parameter. For each table in this paper, the largest numbers in the column are shown in bold.

Table 1. The mean and standard deviation of test accuracy for different distillation algorithms.

Method + Teacher	Accuracy	Params	GPU (MiB)
GM + DD ($K = 60, \zeta_\theta = 50$)	**94.9 ± 0.1**	78.4 K	≈2390
Unroll + DD ($ic = 1$)	88.4 ± 0.3	78.4 K	≈4432
Unroll + DD ($ic = 10$)	79.2 ± 0.7	784 K	≈4426
Unroll + GTN-lrn ($ic = 1$)	92.0 ± 0.3	1.646 M	≈4480
Unroll + GTN-lrn	91.6 ± 0.5 ($ic = 10$)	**1.704 M**	≈4480
Unroll + GTN-rnd	91.7 ± 0.3	1.640 M	≈4480

Table 1 shows the mean and standard deviation of test accuracy, reached by students trained on distilled data. Note that there is only one difference from previous works: we use time limit for each distillation procedure, so there is a degradation in performance. For this experiment, we use $K = 1000, N = 10$ as default hyperparameters values. To check the memory consumption we use a special tool,[5] which can measure the GPU memory usage. Note that using of the **unroll** distillation procedure consumes memory the most. The third column shows the number of teacher parameters, and although **GTN** ($k = 64$) is twice as large as **DD**, there is not much difference in memory usage.

---

[4] https://openreview.net/forum?id=HJg_ECEKDr.

[5] https://pytorch.org/docs/stable/cuda.html#torch.cuda.max_memory_reserved.

6.2   Training Generator with Gradient Matching

In this section we explore the use of the gradient matching to train the teacher generator. We first check the hyperparameters for this distillation method. N controls the frequency of the student's reinitialization, ζ_θ controls the speed at which the teacher's parameters are updated. Figure 2 (a–d) shows the non-trivial relationship between performance and the hyperparameter choice. We assume that such a dependence can be caused by the time limit and the fact that increasing the values of these hyperparameters may cause longer convergence. Note that in previous works [1,3,4] where no time limit was used, increasing ipc always resulted in better performance.

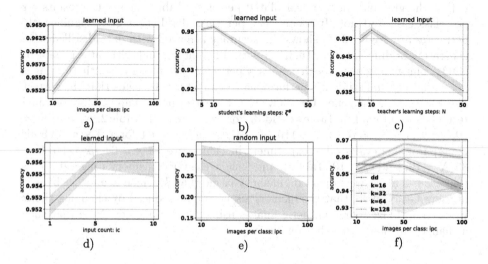

Fig. 2. Dependence of student's performance and hyperparameters of distillation procedure. Next parameters used as default: $ipc = 10, ic = 1, N = 10, \zeta_\theta = 10, k = 64$.

Table 2. Mean and standard deviation of test accuracy for different distillation algorithms.

Method + Teacher	Accuracy	Params	GPU (MiB)
GM + DD	**95.6 ± 0.1**	78.4 K	≈2390
GM + DD (not per class)	86.9 ± 1.5	78.4 K	≈2370
GM + GTN-lrn	**95.2 ± 0.1**	1.646 M	≈2454
GM + GTN-lrn (not per class)	93.4 ± 0.3	1.646 M	≈2434

Figure 2 (e) shows that the fixation of the generator input is really important for gradient matching distillation because teacher training (optimization of λ) diverges when using random input. Another important aspect mentioned above

Table 3. Mean and standard deviation of test accuracy for different distillation algorithms.

Method + Teacher	Accuracy	Params	GPU (MiB)
GM + GTN-lrn ($k = 16, ipc = 100$)	94.2 ± 0.4	172.2 K	$\approx$**4192**
GM + GTN-lrn ($k = 32, K = 50$)	95.9 ± 0.2	449.7 K	$\approx$3610
GM + GTN-lrn ($K = 50$)	96.4 ± 0.1	1.672 M	$\approx$3640
GM + GTN-lrn ($k = 128, K = 50$)	**96.8 $\pm$ 0.1**	**6.533 M**	$\approx$3770
GM + GTN-rnd ($ipc = 10, K = 110$)	29.0 ± 6.1	1.640 M	$\approx$2454

is that the gradient must be calculated per class. Table 2 shows the results for per class case and not. It seems that per class distillation gives significantly better results. Figure 2 (f) shows the accuracy achieved with data distilled with generators of different sizes (marked with different k), and without a generator (**DD**). This plot depicts the dependency between the number of synthetic images per class (ipc) and student's performance on a test set. It seems that the correct size selection for the generator allows to get a better performance. More detailed results can be found in Tables 2 and 3. For experiment in Table 2, we use $ipc = 10$, $ic = 1$, $N = 10$, $K = 110$, $\zeta_\theta = 10$ and $k = 64$ for GTN as default hyperparameters values. For experiment in Table 3, we use $k = 64, ipc = 50, K = 35, N = 10$, and $\zeta_\theta = 10$. Tables 2 and 3 show the GPU memory usage. It seems that ipc has a greater impact on memory usage than k, which is another benefit of using **GTN**. Note that the memory usage can be reduced by changing the ic value to optimize more synthetic images using smaller batches. Note that such a change can slow down the convergence.

6.3 Distillation with Implicit Differentiation

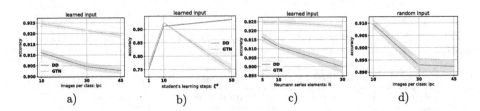

Fig. 3. The relation of the distillation method's hyperparameters and test performance. We use as default: $ipc = 10, N = 10, \zeta_\theta = 10$, and $k = 64$.

The method was proposed in [2], and we will abbreviate it as **IFT** (Implicit Function Theorem). As mentioned above (see Sect. 3), there is no detailed description of the results in the original paper, so they can be found in this section. Figure 3 (a–c) shows the relationship between the hyperparameters of the distillation method and the student's performance on the test. We assume that these results can be explained by the fact that increasing the values of these hyperparameters decreases the frequency of λ update, which negatively affects the performance. The only exception is ζ_θ.

Figure 3 (d) shows results for distillation using a generator with the random input (**GTN-rnd**). Such a generator can produce as much data as we need, but it can not converge when trained with gradient matching. It seems that such distillation becomes possible using implicit differentiation.

Table 4 shows the best results for each method. For this experiment, we use $K = 1080, \zeta_\theta = 50, ipc = 10$, and $N = 10$ as default hyperparameters values. The performance seems to be the same or even better compared to backpropagation through the training procedure **unroll** (see Table 1). Note the difference in memory usage in both tables. Also note that the implicit differentiation distillation is inferior to the gradient matching distillation.

We think this may be connected with the difference in the frequency of λ update. To do one update using **IFT**, we first have to train the student, which is not needed in case of **GM**. It is also important to note that this method is very sensitive to α and ζ_θ, and in some **DD** cases it starts to diverge after several iterations. Meanwhile the use of **GTN** makes the procedure more stable and allows for a more generalizable dataset (see Table 6).

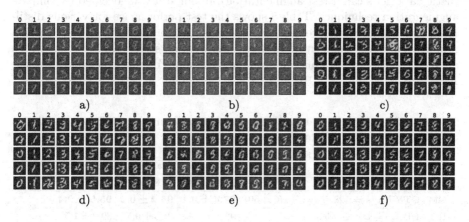

Fig. 4. Synthetic images for MNIST classification task obtained with different distillation methods: a) GM+DD, b) IFT+DD, c) GM+GTN-lrn, d) IFT+GTN-lrn, e) GM+GTN-rnd, f) IFT+GTN-rnd. We use the same hyperparameters as mentioned in Table 5. Hyperparameters for GM+GTN-rnd are described in caption of Table 4.

Table 4. Mean and standard deviation of test accuracy for different distillation algorithms.

Method + Teacher	Accuracy	Params	GPU (MiB)
IFT + DD ($K = 500$)	**93.5 ± 0.5**	78.4 K	≈2726
IFT + GTN-lrn ($\zeta_\theta = 10$)	92.4 ± 0.2	**1.646 M**	≈2726
IFT + GTN-rnd ($\zeta_\theta = 10$)	90.9 ± 0.3	1.640 M	≈2726

Figure 4 shows part of the final synthetic dataset for **GM** (see a, c and e) and **IFT** (see b, d and f). The greatest difference is obtained when data distilled without a generator (see a, b). Synthetic data obtained using implicit differentiation looks less realistic and therefore can be used for federative learning [13]. Also note that the images distilled using a generator are more contrast.

6.4 Distillation with Augmentation

In previous works, augmentation has been used in different ways. In [4] it takes place during distillation (let's call it train augmentation) by applying transformations to real images $\mathcal{B}^T$. In [1,3] it is used when teaching student on synthetic data (let's call it test augmentation). In our study, we decided to compare augmentation techniques. Table 5 shows the test performance for various distillation and augmentation techniques. It seems that for the MNIST classification problem only test augmentation gives improvements (see Tables 2, 3, 4). To augment images we use random crop and rotation. For this experiment, we use $K = 1080, ipc = 10, \zeta_\theta = 10$, and $N = 10$ as default hyperparameters values.

Table 5. The mean and standard deviation of test accuracy for different distillation algorithms and different augmentations.

Method + Teacher	Test Aug.	Train Aug.	Test + Train Aug.
GM+DD ($ic = 1, K = 110$)	96.1 ± 0.4	94.8 ± 0.1	93.9 ± 0.5
GM+GTN-lrn ($k = 128, ipc = 50, K = 50$)	**97.4 ± 0.1**	**96.2 ± 0.2**	**95.5 ± 0.4**
IFT+DD ($\zeta_\theta = 50, K = 500$)	92.3 ± 0.9	91.4 ± 0.5	89.2 ± 1.5
IFT+GTN-lrn	93.0 ± 0.2	91.4 ± 0.3	91.4 ± 0.4
IFT+GTN-rnd	92.2 ± 0.3	89.7 ± 0.3	90.9 ± 0.6

6.5 Generalizability

The generalization problem of distilled data was first mentioned in [1] and then studied in [4] and [7].

Table 6. The mean and standard deviation of test accuracy for different distillation algorithms and student's architectures.

Method + Teacher	LeNet	AlexNet	VGG11	MLP
GM+DD	94.1 ± 0.6	95.0 ± 0.2	95.8 ± 0.3	$\mathbf{88.6 \pm 0.4}$
GM+GTN-lrn	$\mathbf{95.5 \pm 0.3}$	$\mathbf{96.7 \pm 0.2}$	$\mathbf{97.4 \pm 0.1}$	86.8 ± 0.3
IFT+DD	74.0 ± 7.8	68.6 ± 8.9	86.5 ± 1.6	50.9 ± 8.3
IFT+GTN-lrn	91.5 ± 1.0	82.5 ± 14.9	93.0 ± 0.4	79.9 ± 0.6
IFT+GTN-rnd	88.3 ± 2.3	85.3 ± 3.9	92.1 ± 0.4	74.4 ± 1.1

The problem is that such data can't guarantee convergence for students which didn't participate in the distillation procedure. And this problem is of great importance, since the main practical use of synthetic data is the NAS. For this experiment, we use $K = 1080, ipc = 10$, $\zeta_\theta = 10$, and $N = 10$ as default hyperparameters values. Table 6 shows the results of students with different architectures trained on data distilled with different methods. For distillation we used ConvNet student's architecture, all results were obtained with test augmentation. It seems that the best generalizability can be obtained using **GTN** and **GM**. For a comparison with ConvNet see the second column of Table 5.

7   Conclusion

This work explores all the latest ideas in dataset distillation field suggested in [1–4]. We honestly compared the performance of all known methods, limiting their running time. We also proposed new methods based on the joint use of generators and memory efficient methods. Experiments with the MNIST benchmark show that selecting the correct size for the generator allows to achieve better performance for gradient matching distillation, and improves the generalizability of implicit differentiation distillation. This paper also presents the results of augmentation impact on distillation. We also provide a detailed description of the experimental results for implicit differentiation distillation, as we could not find them in the original work [2]. As future work, we would like to experiment with much more diverse datasets and architectures. We also want to improve the distilled data generalizing ability using stochastic depth networks [11]. We are also interested in experiments with bringing the distribution of synthetic objects closer to the original one.

Acknowledgments. This research has been supported by the Interdisciplinary Scientific and Educational School of Moscow University "Brain, Cognitive Systems, Artificial Intelligence".

References

1. Wang, T., Zhu, J., Torralba, A., Efros, A.A.: Dataset distillation. CoRR arXiv:1811.10959 (2018)

2. Lorraine, J., Vicol, P., Duvenaud, D.: Optimizing millions of hyperparameters by implicit differentiation. CoRR arXiv:1911.02590 (2019)
3. Zhao, B., Mopuri, K.R., Bilen, H.: Dataset condensation with gradient matching. CoRR arXiv:2006.05929 (2020)
4. Such, F.P., Rawal, A., Lehman, J., Stanley, K.O., Clune, J.: Generative teaching networks: accelerating neural architecture search by learning to generate synthetic training data. CoRR arXiv:1912.07768 (2019)
5. Maclaurin, D., Duvenaud, D., Adams, R.: Gradient-based hyperparameter optimization through reversible learning. CoRR arXiv:1502.03492 (2015)
6. Sucholutsky, I., Schonlau, M.: Soft-label dataset distillation and text dataset distillation. CoRR arXiv:1910.02551 (2019)
7. Medvedev, D., D'yakonov, A.: New properties of the data distillation method when working with tabular data. In: van der Aalst, W.M.P., et al. (eds.) AIST 2020. LNCS, vol. 12602, pp. 379–390. Springer, Cham (2021). https://doi.org/10.1007/978-3-030-72610-2_29
8. Polyak, B.: Some methods of speeding up the convergence of iteration methods. USSR Comput. Math. Math. Phys. **4**, 1–17 (1964)
9. MNIST Handwritten Digit Database. http://yann.lecun.com/exdb/mnist/. Accessed 17 Apr 2021
10. Grefenstette, E., et al.: Generalized inner loop meta-learning. CoRR arXiv:1910.01727 (2019)
11. Huang, G., Sun, Y., Liu, Z., Sedra, D., Weinberger K.: Deep networks with stochastic depth. CoRR arXiv:1603.09382 (2016)
12. Gidaris, S., Komodakis, N.: Dynamic few-shot visual learning without forgetting. CoRR arXiv:1804.09458 (2018)
13. Zhou, Y., Pu, G., Ma, X., Li, X., Wu, D.: Distilled one-shot federated learning. CoRR arXiv:2009.07999 (2020)

Social Network Analysis

Agent-Based Model for Estimation of Collective Emotions in Social Networks

Kirill Polevoda[✉], Dmitriy Tsarev, and Anatoliy Surikov

ITMO University, St. Petersburg, Russia
polevoda.kirill@mail.ru

Abstract. Emotion spreading in online communities has drawn the attention of social sciences for decades; an actual problem in this area is estimating collective emotions in social networks (Twitter, Facebook, Weibo). People chat on community walls exchanging messages transmitting and receiving emotions; we describe these emotions in terms of Ekman's six-emotion model: Joy, Sadness, Surprise, Anger, Fear, and Disgust. The sum of all the messages makes an emotion field or collective emotions, which we aim to measure. We propose an agent-based model: we generate a number of agents that simulate human perception of emotions. The agents monitor the community, "read" the posts and comments on the wall, estimate the emotions of each read text, taking into account not only the words but the emojis end emoticons as well. Then every agent makes a prediction for collective emotions of the community in the moment of time. Since the agents differ by their parameters, the average value of their estimations represents an "objective" measure of the emotion field as it is perceived by users. For the emotion estimation in a particular text agents use a simple sentiment analysis based on emotional dictionary. However, this approach is sufficient for our purposes, we discuss the possibility to further improve the prediction quality via Bayesian network approach. A naive Bayesian network has already been implemented and demonstrated a better prediction; this approach provides tools to extract significantly more information about emotions from the natural language text then the sentiment analysis. The results of social experiment with actual users demonstrate a good agreement of the agent-model prediction and the human evaluation of collective emotions. The results obtained may find application in social and political sciences, marketing, and target commercial.

Keywords: Agent model · Emotional field · Emotion recognition

1 Introduction

Information spreading in social networks is one of the actual topics in modern computer and social sciences. Providing fast communication of large groups of people, social networks intensively affect the spread of opinions, rumors, consumer preferences, etc. [1, 2]. Especially, the dynamics of emotion transition between the users attracts the attention of scientist all around the world; the results of such studies may be applied in marketing [3–6], political [7, 8] and social sciences [9, 11].

© The Author(s), under exclusive license to Springer Nature Switzerland AG 2022
E. Burnaev et al. (Eds.): AIST 2021, CCIS 1573, pp. 153–165, 2022.
https://doi.org/10.1007/978-3-031-15168-2_13

One special form of communication in social networks is chatting on communities' walls. When many users post messages publicly, it is impossible for a person to read all of them. In this case, when users responds some post, no one can tell (even the user themself), which messages have affected his opinion and mood. For a researcher studying the emotion dynamics in a social network this situation evokes difficulties. A possible solution is to consider an emotion field, or collective emotions, in a community, somehow averaging all of the users' emotions. However, direct averaging is not the best option. As known, the emotion perception is subjective [10], and this must be taken into account. Moreover, as pointed before, a rare user reads all of the posts on the wall, which contradicts the idea of direct averaging.

In this work we propose an agent-based model for emotion field estimation in a social network community. Small digital entities, which we call agents, simulate the users, who read posts and comments on a community wall, measure the emotions of these messages, and make an estimation of the emotion field in the moment of time and average emotion field of the community. Then, averaging the estimations of a number of agents with different parameters allows to obtain a measure of emotions (in the form of time series), which one can call objective. Thus, our approach combines the subjectivity of estimations of the particular agents (users) and the objectivity of estimation made by the whole community, providing the "inside view" on the collective emotions of the community. Notice, the agent-based models are popular for describing the emotion dynamics in networks [12, 14, 15]. However, the agent-based approach to the emotion estimation itself seems to be of rare usage, which makes the scientific novelty of the paper.

The paper is arranged as follows. In Sect. 2 we provide a short overview on the state of the art of the emotion field estimation. Section 3 is devoted to the methods we use for the emotion estimation in the text; the proposed agent-based model is described in detail in Sect. 3 as well. In Sect. 4 we describe the social experiment on emotion field estimation, which we have performed to validate the agent-based prediction. Section 5 contains the main results of the emotional field estimation in two online social communities under consideration; we discuss these results and their possible implementation in Sect. 6. A short summary is presented in Conclusion.

2   Literature Overview

As mentioned, the agent-based models are quite popular for describing the emotion dynamics in social communities. For example, in [12] the authors propose such a model to explain the emergence of collective opinions in social communities. They use a simple 2D model describing human emotions in terms of Valence and Arousal [13]. The authors also use the concept of emotional field generated by the agents with high arousal; it affects the valence and arousal of all other agents. In this model the emotion field is just a sum of all positive- and negative-valence emotions of all the actors; it changes in time according to agents' activities. This model does not account the difference between such emotions as Joy and Surprise or Sadness and Disgust, and therefor it is not capable to describe agents' different responses on different emotions of the similar valance. However, the authors derive the critical conditions for emotional interactions to obtain either consensus or polarization of opinions.

In [14] the agent-based approach is established to formalize and simulate emotion contagion processes within groups, which may involve absorption or amplification of emotions. As the authors claim, the type of emotion is not specified, in principle it can be any emotion, for example six Ekman's emotions that we consider. Similarly, in [15] the agent-based model is proposed for emotion contagion and competition in online social media. The agents of the model simulate the behaviour of people communicating in Weibo, Twitter or similar social networks. Based on 11 million tweets from Weibo, the model effectively reproduces the empirical patterns of transmission such emotions as Anger, Joy, Disgust and Sadness. But again this model does not include any emotion field. On the other hand, the authors of [23] introduce a framework to link collective emotions, which can be described as emotional field, with emotions of individual users. This model can be used for agent-based modeling of collective emotions for social communities. In [24] the authors show the influence of collective emotions on the inter-action between users in Internet communities and on trajectories of the communities' existence.

In [22] the authors suggest using the agent-based approach to model the emotions of every node of Collaborative Network (CN), which includes information about com-panies, to better understand the behavioral dynamics of the CN in virtual environment. The authors describe the model as a set of agents that individually interact with the environment in different ways, depending on their parameters.

Many papers focus on collective emotions describing the evolution of the emo-tions of every particular user and then simply averaging them [17–19]. The approach to collective emotions as average values is simple, obvious, and good enough for many purposes, for example for collective decision-making tasks [20,21]. However, such an approach does not account for the subjectivity of the emotion perception and also is not applicable, when the amount of information is too large for a person to fully perceive. Our global research (beyond the current paper) is devoted to the emotional dynamics in online social communities, so the subjectivity of the emotion perception and the selec-tivity of user's posts reading are in the core of it. Thus in this paper we develop an agent-based approach focusing on these issues.

Talking about the emotion field estimation we should mention the approaches of the emotion detection in a given text. There are several popular approaches for that, namely the convolutions network [26], rule-based [28], natural language preprocessing [29], and sentiment analysis [30] approaches. We use the approach similar to the one in [30], a sentiment analysis based on emotional dictionary, as it is simple to perform and provide a decent quality of emotion estimation.

3 Methods

3.1 Estimation the Emotions in One Text

For the text sentiment analysis, we use an emotional dictionary of the Russian lan-guage [31]. In the dictionary every word is linked to the six-dimensional vector denoting the emotions, carried by the word, according to Ekman's model. Thus, when analyzing the text for each word the vector is determined as

$$a_j = \{e^{(i)}\}, \quad i = 1..6, \tag{1}$$

where e indicate the presence or absence of i emotion in the word, could be "1" and "0" respectively.

To estimate the emotions of the text, first, all the text words are lemmatized, the emoticons and emojis are turned into their text equivalents from the dictionary. Then all vectors (1), corresponding to the preprocessed text, are summed, and the result is a vector of the text dominant emotions

$$a = \sum_{j=1}^{D} a_j, \tag{2}$$

where D is the number of words in the text.

Finally, vector (2) is converted to five-point scalars for each emotion i according to the following rule:

- 0 if there is no emotion i in the text
- 1 if emotion i are less than 15% of D
- 2 if emotion i are less than 30% of D
- 3 if emotion i are less than 45% of D
- 4 if emotion i are more than 45% of D

As a result, for each text a six-dimensional five-point vector of emotions is obtained. The database for the model approbation is based on communities "CS Tyumeni" ("Exidents of Tymen city") and "Tonkiy yumor" ("Inside jokes") in Facebook-like Russian social network "VK". We have collected 180 posts (120 for the first community and 60 for the second one), with all the comments, for year 2019. Notice, the study of "Tonkiy yumor" is in progress now, so only data for a half of year 2019 is processed. In total 9938 texts were used for the model calibration and validation.

3.2   Agent-Based Model for Collective Emotions Estimation

We present an agent-based model for estimation of collective emotions in social network communities. For our modeling we use $Mesa$ – an Apache2 licensed agent-based modeling (or ABM) framework in Python.

The agents are small digital entities "reading" the communities' walls and estimating the collective emotions. The agents differ in several parameters:

- N and n – The maximum number of posts and comments that an agent reads being activated.
- α – **Inertia** – The degree agent's memory; it reflects how the previous activation results influences the current assessment.
- β – **Susceptibility** – The agent's focus when determining the collective emotions of the community: on particular bright emotional texts or on the average mood of the community.
- γ – **Attentiveness** – The factor of randomness indicating whether the agent skips some posts and comments.

The algorithm of the collective emotions estimation is described in Fig. 1. The model activates all the agents simultaneously at regular intervals of time. When activated, the agent estimates the instantaneous value of the collective emotions $G = \{G^{(i)}\}$, $i = 1..6$ after reading all posts and comments as

$$G^{(i)} = \beta \cdot \max_{k=1..P} \left[a_k^{(i)} \right] + (1 - \beta) \cdot \sum_{k=1}^{P} \frac{a_k^{(i)}}{P}, \tag{3}$$

where P is the number of posts and comments read, $a_k^{(i)}$, $k = 1..P$ are five-point scalars for i-th emotion of all read texts. All agents' estimations G are stored in the model memory for the further analysis.

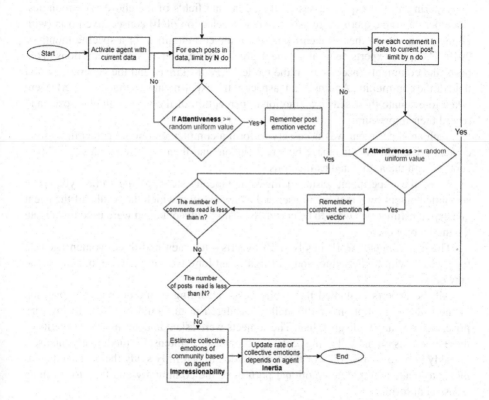

Fig. 1. Algorithm of collective emotions estimation by agent when activated.

The agent updates the collective emotion score G_{cur} based on the score at the previous point in time G_{rec} as

$$G_{cur} = \alpha \cdot G_{rec} + (1 - \alpha)\, G. \tag{4}$$

Then the agents fall asleep until the next activation after a period of time Δt. To estimate the collective emotions we average all the agents' estimations as

$$G_{av}(t) = \frac{1}{M} \cdot \sum_{k=1}^{M} G_k(t), \tag{5}$$

which represents an emotional field of the community in the moment of time.

4   Online Psychological Experiment

To test the above model, a cyber psychological experiment has been performed. During the experiment, the experts assessed the emotional fields of the chosen communities. Once per day the experts were asked to read a selection of 10 consecutive posts (with all the comments to them) taken from one of the communities for a specific month of 2019. Then, the experts determined the degree of expression of specific emotions in the posts and comments (according to the model of Paul Ekman) and the general emotion field of the community. An important aspect of the experiment was that the experts were asked to evaluate their sensory, emotional, perception of the viewed public posts, and record their impression.

Such an experiment was necessary to form a complex emotional perception of the content presented in cyberspace by a real person, and then to correlate this "live" perception with the automated estimations

Thus, the experiment assumed the simultaneous study of the same cyberspace semantic content by a real living user and an agent, after which the results of the agent and expert estimations were to be correlated. The results obtained were used to validate the model proposed.

The experimental work involved 13 experts - ten men and three women aged 22 to 39 years, with higher education and significant long-term experience in using social networks.

All the experts evaluated the public posts according to the degree of expression of the following emotions traditionally considered in this kind of research: joy, surprise, sadness, anger, disgust, fear. The subjects were asked to determine the severity of these emotions on the following scale: "very intense", "intense", "moderately intense", "weakly intense", "absent". For that they needed to carefully study the posts and comments to make a complete emotional portrait of the community and then to fill in an assessment form.

In addition, the experts were also asked to fill in a form reflecting their impression of the community's general emotion field according to the following scale: "very positive", "mostly positive", "neutral", "mostly negative", "extremely negative".

5   Results

Using the markup, provided by 13 experts, we have validated our model with ROC-AUC and Precision-Recall-Curve-AUC scores. All curves for each emotion are shown in Figs. 2 and 3. The curves and AUC-scores demonstrate how well our model estimates the collective emotions based on experts' markup. The area under the curve of the ideal model is 1. Due to the fact that the emotion estimations are five-pointed vectors, a threshold of "2" was applied for the experts'a marking, returning "1" if the expert's emotion value is greater or equal, and "0" otherwise. The normalized agent-based estimations, depending on the agents parameters, were evenly distributed from "0" to "1".

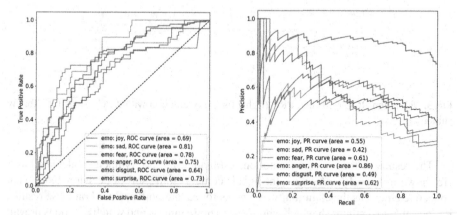

Fig. 2. Validation ROC (left) and PRC (right) for every emotion with AUC values for "CS Tyumeni"

Thereby, the curves were constructed; For the community "Tonkiy yumor" cu. AUC-scores demonstrate a good agreement with the expert assessment. We have also calculated precision, recall, accuracy, f1-score for each emotion, these metrics are presented in Table 1.

Table 1. Validation evaluated metrics

Group	Metric	Joy	Sad	Fear	Anger	Disgust	Surprise	Mean
CS Tyumeni	Precision	0.564	0.24	1.0	0.891	0.44	0.75	0.648
	Recall	0.657	1.0	0.068	0.395	0.183	0.194	0.416
	Accuracy	0.661	0.452	0.756	0.518	0.625	0.679	0.615
	f1-score	0.607	0.387	0.128	0.547	0.259	0.308	0.373
Tonkiy yumor	Precision	1.0	0.24	0.872	1.0	1.0	0.761	0.812
	Recall	0.242	0.261	1.0	0.2	0.245	1.0	0.491
	Accuracy	0.372	0.8	0.872	0.711	0.794	0.761	0.718
	f1-score	0.389	0.25	0.932	0.333	0.393	0.864	0.527

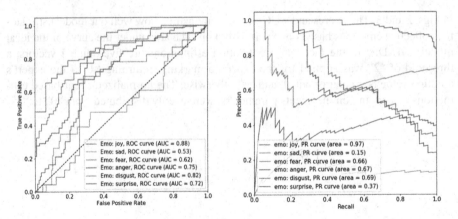

Fig. 3. Validation ROC (left) and PRC (right) for every emotion with AUC values for "Tonkiy yumor"

The results of the collective emotions estimation in "CS Tyumeni" calculated as 5 are shown in Figs. 4(b). As seen, the agents come to some sort of decision about what collective emotions the community expresses for each emotion. Based on these results, it is possible to distinguish which emotions are more intense and which emotions prevail in the community. Notice, the most intensive emotion is sadness, as well as the other negative emotions are more intensive than the positive ones. The emotion of joy is manifested in most of the posts and comments, this is the reason of its vivid predominance in the group.

The model validation on community "Tonkiy yumor" data is in progress. However, current data already shows good validation results for all the emotions.

6   Discussion

6.1   Psychological Patterns of the Communities

In this section we briefly analyse the results of emotion estimation in two online communities from the position of psychology. The results of emotional fields estimation are shown in Figs. 4.

As seen from Figs. 4 the emotional field estimations made by agents after some period of growing saturate and stabilize. The expert markups obtained during the social experiment confirm this result demonstrating the saturation as well. This saturation may be explained by memory effect, when the social network users following the community become biased towards it. On large time scales the estimations of the emotional field depend more on the previous experience and less on the new data; this is valid both for the human respondents and digital agents, as the social experiment demonstrate.

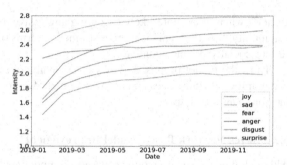

(b) Mean estimation of all agents in the model for every emotion for "CS Tyumeni"

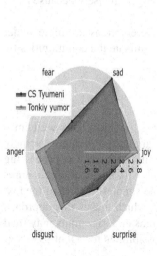

(a) Difference in emotion intensity for last timestamp

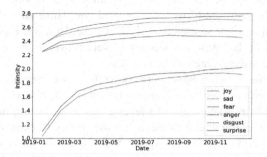

(c) Mean estimation of all agents in the model for every emotion for "Tonkiy yumor"

Fig. 4. Mean estimation for both groups

Another hypothesis here is that the emotional field of the community behave periodically within some time intervals, for example during one year as the most of life processes (and person traits) depend on season. Thus when the observation of the emotional fields is continued, we expect some periodic functions instead of saturation. Testing this hypothesis may be the basis for a separate study.

Analyzing Figs. 4, especially Fig. 4(a), one can see the differences in the communities' emotions are not significant but evident. The community "CS Tyumeni" demonstrates more pronounced fear, surprise, and sad, while joy, anger, and disgust are more relevant to "Tonkiy yumor". The community "CS Tyumeni" is socially significant accumulating important information related to the life quality in a certain area including the accident reports. On the other hand, the community "Tonkiy yumor" is more entertaining, containing recreational content. Thus the emotional reaction of the users on the unpleasant content in these communities is different: it more often causes sadness in "CS Tyumeni" but disgust in "Tonkiy yumor". The emotion anger is produced mostly by the users' comments, not content itself. As the content analysis demonstrated, aggressive, provocative comments arise more often in the entertainment community "Tonkiy yumor", where they more likely case long aggressive discussions. We assume that the community "CS Tyumeni" is more restrained in terms of anger precisely because of its serious socially-oriented content.

Thus, the model proposed allows to estimate collective emotions in online social communities as time-series and as average metrics to discriminate the communities by the dominant emotions.

6.2 Bayes Network Based Model Improvement

Although the emotional dictionary based sentiment analysis provide a quite well emotion estimation in text, we must discuss a better method familiar in literature. Namely the Bayes network approach is worth mention. A Bayes network for text processing in terms of emotion recognition represents a directed graph with nodes of text features connected with six nodes of Ekkman's basic emotions. The simplest feature of the text is the presence of words associated with some emotions, which makes this approach similar to the sentiment analysis we use. However bigramms and trigramms may (and should) be taken into account as well, providing more possibilities for agent to recognize and estimate emotions in text.

We train our Bayes network on the set of texts marked by experts as containing one or several emotions (or no emotions at all), such as joy, sadness, surprise, anger, disgust, and fear. When the model faces a new token (word, bigramm or trigramm) in the text used for training, it creates a new token node and connect the nodes of emotion classes, marked in the text by experts, with the new token node by the link of weight "1". If the faced token is familiar for the model, and the correspondent node already exists, then the model creates new emotion-token links and re-count the link weights as n_i/n, where n_i is the number of times, when the token was found in the texts with the i-th emotion, and n is the total number of the token appearance in texts, including the ones with no emotions.

At the output we have a directed weighted graph capable for agents to exploit in emotion recognition. When the agent analyzes some text, the feature (token) nodes activate and then they activate the emotion nodes, providing the possibility of the basic emotions present in the text. This approach is similar to the Naive Bayes Classifier (NBC) [32] widely used for information retrieval and data analysis, including the emotion detection [33,34]. It is called naive because it is assumed that the text features, such as the appearance of words, bigramms, and trigramms, are independent (which is naive). The further improvement here is directed on accounting more complicated features including cross-dependence of them. We expect the improvement of our agent-based model for collective emotions estimation in turns of emotion sensitivity and prediction accuracy.

6.3 Model Implementation

The agent-based model were developed for the emotion dynamics problem in social network communities, which is to be solved further. Strictly speaking it requires us to improve our agents, make them capable not just precept the emotions but also react on them. We describe the emotion dynamics of the social network users with the simple independent cascade model. We assume that the user reading the community wall precept its emotional field and the emotions it contains leads to the users' emotional state changing. Moreover, we assume that all the six field emotions are connected and each of the emotion can more or less create all them. The ratios of this correlations make the emotion correlation matrix, which we assume is unique for different persons and stable at the times of observation. Thus, observing the user's digital footprints and estimating the collective emotions in the community we wish to analyse the user's personality. This can be applied for the recognition of bots and fake accounts, as well as for distant diagnostics of psychological states of users.

7 Conclusion

To summarize, we propose the agent-based model for the collective emotions (emotional field) estimation in online social communities. The ROC and AUC curves, presented in Figs. 2 and 3, demonstrate good results. The curves show how well the real data is modeled, the area under the curve shows the modeling quality. The values greater than 0.7 are good for simulating emotions. For almost all emotions, the area is greater than 0.7, which indicates good validation results. Also, the average validation accuracy for all emotions is 61.5 for the first community 1 and 76.3 for the second one. This value indicates the correctness of the collective emotions assessment by the agent. Thus, the model demonstrates its validity and suitability for implementation. The strong point of the approach proposed is the equipment the agents with subjectivity combined with the objectivity provided with their amount. Simulating the users behaviour the agents provide an inside collective emotions estimation, allowing to see the community from the users' point of view.

In this paper we have proposed a novel model of the agent-based collective emotion estimation. The main feature of the proposed model is the subjective side of emotion recognition. The model allows to simulate the features of real people, making the emotion assessment more accurate and more human-like.

The model developed is to be used in the future to predict and simulate the spread of emotions in social networks, which is the topic of our further research. We also plan to further improve the model quality of estimation.

References

1. Katona, Z., Zubcsek, P.P., Sarvary, M.: Network effects and personal influences: the diffusion of an online social network. J. Mark. Res. **48**(3), 425–443 (2011)
2. Bell, D.R., Song, S.: Neighborhood effects and trial on the Internet: evidence from online grocery retailing. Quant. Mark. Econ. **5**(4), 361–400 (2007)
3. Schaat, S., Wilker, S., Miladinovic, A., Dickert, S., Geveze, E., Gruber, V.: Modelling emotion and social norms for consumer simulations exemplified in social media. In: 2015 International Conference on Affective Computing and Intelligent Interaction, (ACII), pp. 851–856. IEEE (2015)
4. Leitch, K., Duncan, S., O'Keefe, S., Rudd, R., Gallagher, D.: Characterizing consumer emotional response to sweeteners using an emotion terminology questionnaire and facial expression analysis. Food Res. Int. **76**, 283–292 (2015)

5. Tsai, W.-C., Huang, Y.-M.: Mechanisms linking employee affective delivery and customer behavioral intentions. J. Appl. Phys. **87**(5), 1001 (2002)
6. Berg, H., Söderlund, M., Lindström, A.: Spreading joy: examining the effects of smiling models on consumer joy and attitudes. J. Consum. Mark. **32**(6), 459–469 (2015)
7. Khrennikov, A.: "Social Laser": action amplification by stimulated emission of social energy. Phil. Trans. R. Soc. A **374**, 20150094 (2016)
8. Grover, P., et al.: Polarization and acculturation in US Election 2016 outcomes-Can twitter analytics predict changes in voting preferences. Technol. Forecast. Soc. Chang. **145**, 438–460 (2019)
9. Khrennikov, A.: Information Dynamics in Cognitive, Psychological, Social, and Anomalous Phenomena. Fundamental Theories of Physics. Kluwer Academic Publishers, Dordrecht (2004)
10. Barrett, L.F., Lewis, M., Haviland-Jones, J.M.: Handbook of Emotions, 4th edn, p. 928. Guilford Publications (2016)
11. Burke, M., Marlow, C., Lento, T.: Social network activity and social well-being. In: Proceedings of CHI 2010, pp. 1909–1912. ACM Press (2010)
12. Schweitzer, F., Krivachy, T., Garcia, D.: An agent-based model of opinion polarization driven by emotions. Complexity (2020)
13. Schweitzer, F., Garcia, D.: An agent-based model of collective emotions in online communities. Eur. Phys. J. B **77**(4), 533–545 (2010)
14. Bosse, T., et al.: Agent-based modeling of emotion contagion in groups. Cogn. Comput. **7**(1), 111–136 (2015)
15. Fan, R., Xu, K., Zhao, J.: An agent-based model for emotion contagion and competition in online social media. Phys. A **495**, 245–259 (2018)
16. Mitrović, M., Tadić, B.: Dynamics of bloggers' communities: bipartite networks from empirical data and agent-based modeling. Phys. A **391**(21), 5264–5278 (2012)
17. Hołyst, J.A., Kacperski, K., Schweitzer, F.: Phase transitions in social impact models of opinion formation. Phys. A **285**(1–2), 199–210 (2000)
18. Hołyst, J.A., Kacperski, K., Schweitzer, F.: Social impact models of opinion dynamics. Ann. Rev. Comput. PhysicsIX, 253–273 (2001)
19. Xiong, X.B., et al.: Dynamic evolution of collective emotions in social networks: a case study of Sina weibo. Sci. China Inf. Sci. **56**(7), 1–18 (2013)
20. Tsarev, D., et al.: Phase transitions, collective emotions and decision-making problem in heterogeneous social systems. Sci. Rep. **9**(1), 1–13 (2019)
21. Khrennikov, A.: 'Social Laser': action amplification by stimulated emission of social energy. Philos. Trans. R. Soc. A Math. Phys. Eng. Sci. **374**(2059), 20150094 (2016)
22. Ferrada, F., Camarinha-Matos, L.M.: A system dynamics and agent-based approach to model emotions in collaborative networks. Technol. Innov. Smart Syst. **499**, 29–43 (2017)
23. Garcia, D., Schweitzer, F.: Modeling online collective emotions. Chair Syst. Des. **37** (2012) https://doi.org/10.1145/2390131.2390147
24. Chmiel, A., Sienkiewicz, J., Thelwall, M., Paltoglou, G., Buckley, K., et al.: Collective emotions online and their influence on community life. PLoS ONE **6**(7), 1–8 (2011)
25. Jin, S., Zafarani, R.: Emotions in social networks: distributions, patterns, and models (2017). https://doi.org/10.1145/3132847.3132932
26. Shrivastava, K., Kumar, S., Jain, D.K.: An effective approach for emotion detection in multimedia text data using sequence based convolutional neural network. Multimedia Tools Appl. **78**(20), 29607–29639 (2019). https://doi.org/10.1007/s11042-019-07813-9
27. Von Scheve, C., Ismer, S.: Towards a theory of collective emotions. Emot. Rev. **5**(4), 406–413 (2013)
28. Shaheen, S., et al.: Emotion recognition from text based on automatically generated rules. In: IEEE International Conference on Data Mining Workshop, pp. 383–392 (2014)

29. Alm, C.O., Roth, D., Sproat, R.: Emotions from text: machine learning for text-based emotion prediction. In: Proceedings of Human Language Technology Conference and Conference on Empirical Methods in Natural Language Processing, pp. 579–586 (2005)
30. Rao, Y., Lei, J., Wenyin, L., Li, Q., Chen, M.: Building emotional dictionary for sentiment analysis of online news. World Wide Web **17**(4), 723–742 (2013). https://doi.org/10.1007/s11280-013-0221-9
31. Surikov, A., Egorova, E.: Emotional analysis of Russian texts using emojis in social networks. In: van der Aalst, W.M.P., et al. (eds.) AIST 2020. LNCS, vol. 12602, pp. 282–293. Springer, Cham (2021). https://doi.org/10.1007/978-3-030-72610-2_21
32. Rish, I., et al.: An empirical study of the Naive Bayes classifier. In: IJCAI 2001 Workshop on Empirical Methods in Artificial Intelligence, pp. 41–46 (2001)
33. Asriadie, M.S., Mubarok, M.S.: Classifying emotion in Twitter using Bayesian network. In: Journal of Physics: Conference Series, p. 012041 (2018)
34. Abbasi, M.M., Beltyukov, A.P.: Analysis of emotions from the text in Russian using syntactic methods. Inf. Technol. Syst., 137–142 (2019)

A Method for Identifying Bridges in Online Social Networks

Andrey N. Rabchevskiy[1,2]([✉]) [iD], Victor S. Zayakin[1,3] [iD],
and Evgeny A. Rabchevskiy[1] [iD]

[1] JSC "SEUSLAB", Perm, Russia
andrey@ranat.ru
[2] Perm State University, Perm, Russia
[3] National Research University "Higher School of Economics", Perm, Russia

Abstract. The current level of development of online social networks has transformed social media from a way of communication between people into a tool for influencing people's behaviour in their daily lives. This influence is often aimed at inciting protest movements in society and mobilising citizens for protest actions, and has a targeted impact on social network users. The sponsors and main actors of disruptive influences are often forces located in other countries. In the context of counteraction to targeted destructive influences, the task of identifying the network structure of destructive influence is very relevant. One element of this structure is the users connecting individual communities to the core of the protest network. These users are the bridges between the clusters and the core network. Their main task is to contribute to the rapid growth of the protest audience. Identifying the most influential bridges and blocking them could decrease the protest potential or make the protest actions ineffective. In this paper, we propose a methodology for identifying bridge users based on the original centrality measure of weighted contribution. Moreover, a method for identifying the most influential bridges is proposed. Unlike most probabilistic methods, weighted contribution centrality allows for clear determination of whether a user is a bridge or not. A description of the measure, a mathematical model and an algorithm for calculating it are presented.

Keywords: Online social networks · Social network analysis · Structure of protest network · Core of protest network · Clusters · Community · Bridges · Weighted contribution centrality

1 Introduction

Today's social networks are no longer just a means of communication between people and have evolved into an effective tool for targeting users. The aim of influence can be to engage users in specific thematic communities or to disseminate information that can influence people's behaviour in everyday life.

© The Author(s), under exclusive license to Springer Nature Switzerland AG 2022
E. Burnaev et al. (Eds.): AIST 2021, CCIS 1573, pp. 166–175, 2022.
https://doi.org/10.1007/978-3-031-15168-2_14

Examples of these influences are the political events of the Arab Spring in 2010–2011, the #Occupay movement in the US in 2011, the protests in Turkey, Brazil and Hong Kong (2013–2014), the recent presidential elections in Belarus (2020) and the political actions around the arrest of Navalny and "Putin's palace" in 2021, where social media were used to coordinate people into actual political actions.

The study of the mechanisms and degree of influence of social networks on people's behaviour has generated a great deal of scientific interest. According to [1–3], all protest movements are inextricably linked to the creation of autonomous communication networks supported by the Internet. The significant impact of social networks on the level of people's mobilization for action has been described in [4–6]. When studying social networks in the context of protest sentiments, one often observes their pronounced cluster structure. Figure 1 shows examples of graphs of such networks, where the vertices of the graph are users and the edges are connections between them. The colours indicates the level of publication activity of users in the social network, i.e. the number of any type of material on the target topic published by the user. Red indicates the maximum level of publication activity and grey indicates no activity.

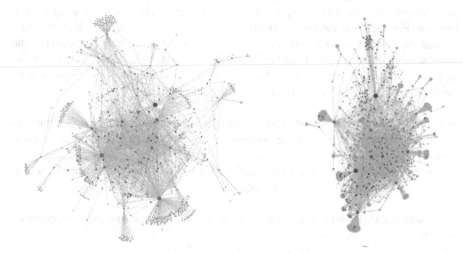

Fig. 1. Examples of the cluster structure of social network graphs.

As can be seen from Fig. 1, most of the graphs have a pronounced cluster structure in the form of a core with many cross-links between users and isolated clusters that are connected to the core through a single user acting as a "bridge" between the cluster and the core.

Since the publication activity of users in clusters is similar to core users, it is logical to assume that the sum of the activity levels of each node in a cluster can be higher than that of any node in the core, and the node connecting the cluster to the core will contribute more to the overall network activity level than any node in the social network core. An analysis of the profiles of users acting as bridges in the protest activity theme showed the following results:

- users in clusters often only partially share the views of core users on the objects of discussion;
- users in clusters are often united by the same topic;
- the preferences, interests, and political views (for protest networks) of users in different clusters may differ (they may belong to different political parties or movements), but these users share opposition to the current authorities;
- as a rule, users connecting the cluster to the core act as community moderators;
- such users have connections with each other and form a substructure in the graph of social relations, due to which they can coordinate their actions, involving in the social phenomenon under study different categories of users, possibly disagreeing with the common point of view of the network core on some issues.

Thus, identifying users who act as social media bridges is crucial to counteracting protest movements and managing the parameters of the spread of viral and destructive information on social media.

Using the software "SEUS search engine" [7], actively used by law enforcement agencies of the Russian Federation [8], we searched for publications in the social network VKontakte related to the organization of protest events in January–August 2019 in Moscow. For each user the level of publication activity was calculated, which took into account the number of posts, reposts, comments, likes, etc. As a result of ranking by the level of publication activity, a ranking of user activity was compiled. For each user the graphs of the social connections of the users' friends and friends of their friends were constructed. The following conditions were taken into account:

- a user is included in the graph if he is a friend of a member of the activity rating or is a friend of any of his friends (the maximum distance to the target user in the graph is two);
- a user whose activity level is zero is included in the graph only if he is a friend of at least two users from the activity rating.

A social network node that satisfies the following requirements was considered a bridge:

- a node that connects the cluster to the core of the network;
- cluster nodes are only connected to the bridge and are not connected to each other;
- bridge is connected to cluster nodes and core nodes.

Figure 2 shows a fragment of a typical node acting as a bridge.

In graph theory these nodes are usually called articulation node, cut-node or broker, but we will use "bridge" for ease of reference. Thus, the challenge was to select or develop a methodology that unambiguously identifies bridges in cluster networks and also determines the extent to which bridges influence the overall level of network activity.

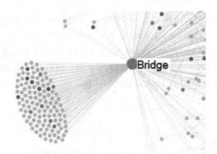

Fig. 2. Fragment of the user-bridge graph.

2 Materials

Various centrality metrics are used to identify the characteristics of nodes in networks, as described in the review paper [9]. The best known centrality measure that characterizes the communicative ability of nodes is the centrality on betweenness, first introduced independently [10] and [11] and finally formulated in [12]. Betweenness corresponds to the sum of all the shortest paths that pass through a given node in the graph. Since each node in the networks we studied had a certain level of activity, it was necessary to consider the weight of each node in the network. To calculate intervening centrality for weighted networks, the techniques proposed in [13–17] could be used. However, betweenness centrality, with or without weight, can reveal the level of communication capability of a node in the network, but cannot accurately determine whether a given node is truly a bridge, since nodes with a high betweenness centrality value can be located both at the core of the network and at the periphery of the network, being bridges.

Influential nodes, according to [18], always act as a "bridge" between communities and exist within an overlapping community. The authors suggest using the local centrality method to identify such influential nodes, which assumes that the more communities a node belongs to, the more influence it has. In [19, 20], "transmission centrality" and "modular centrality" measures are proposed to define bridges, but transmission centrality can be high in both core nodes and bridges, so its meaning is not very different from intermediate centrality, and in modular centrality, nodes connecting communities are the bridge, whereas we investigated nodes between communities and the network core, meaning that the concept of bridge had a different meaning in this context.

A method that successfully identifies bridges is presented in [21], in which the authors introduced the concept of Bridging Centrality. This measure identifies bridges more accurately, but it works only in sparse networks with a large number of bends, because it is based on the idea that to identify bridges it is necessary to discard the value of links with nodes that are in close proximity to a node, that is, links of the first knee of the graph. Since, in our case cluster users are connected only with a bridge, they cannot be taken into account in the calculation of this measure, which did not suit us. The closest measure for our problem is "Contribution centrality" proposed by [22], the essence of which is that the centrality of a node is proportional to the sum of centrality of nodes in its neighborhood, weighted by their contributions. The contribution centrality is indeed the most applicable for our problem, since it can determine the contribution of each bridge for the kernel users, but it does not guarantee an unambiguous definition of the bridge, which in our case was a necessary condition.

As we can see, all the measures presented above could, to a greater or lesser extent, determine the level of communication capability of a node, but cannot exactly determine whether a given node is a bridge as we understand it.

3   Method

We will say that all users with publication activity on the topic of a given social phenomenon and their social connections constitute the "temporary social network" generated by this social phenomenon, and the sum of the activity levels of all users constitutes the total activity level of the temporary social network.

Since the number of users in different clusters and their level of activity are different, bridges can have different levels of influence. Let the degree of influence of the bridge on the overall level of publication activity of the temporary social network be defined as the total level of activity of the cluster that is connected to the core through the bridge. According to the above definition of bridge, cluster nodes should only be connected to the bridge and should not be connected to each other. Consider the graph shown in Fig. 3 and calculate which of the nodes in the graph is a bridge in the context of the proposed definition and calculate its cluster weight.

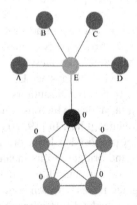

Fig. 3. Node network diagram including weights.

For the red node, links with the green and grey nodes will not be taken into account as they have links with other nodes in the network, so the value of the red node's total rating, as well as for the grey nodes, will be 0.

For blue nodes which only have links with the green node, the value will also be 0 as the green node also has other links,

For the green node the link with the red node will also give 0, and the links with the blue nodes will give the value of the weights of those nodes,

The weight of the green node will be equal to the sum of the weights of the blue nodes $E = A + B + C + D$.

As we can see, we obtained a single non-zero value for the green node in the whole network, which exactly determines the presence of the sought bridge and its contribution to the overall level of activity of the temporal social network, equal to the value of E.

Given that the weight of each node in the graph plays a significant role in the proposed method, the term "Weighted Contribution Centrality" was proposed to determine the degree of influence of the bridge on the overall level of activity.

We will say that weighted contribution centrality is the contribution of a social network node to the total level of publication activity, equal to the sum of the activity of each cluster node connected by the node to the core network, divided by the total activity level of the network. In other words, bridge weighted contribution centrality is the weight of the cluster connected by the bridge to the core, divided by the total weight of the network.

The weighted contribution centrality value $= 0$ if the node is not a bridge, and > 0 if the node is a bridge. The most influential node in the bridge role has the highest weighted contribution centrality value for this network. Let us introduce notations to formally describe the proposed methodology.

Let $G = (U, F)$ be a graph consisting of a set U of users and a set F of disordered pairs of different elements of the set U, reflecting friendly relations between users (graph edges).

If users u and v are friends, i.e. form a relation $f \in F$, we write $f = (uv) = (vu)$. Denote the set of friends of user $u \in U$ by $F(u) = \{v \in U: (uv) \in F\}$. Then the degree of a node, i.e. the number of friends of user $u \in U$, is naturally denoted by $|F(u)|$. The set of users associated only with a given user u is called the neighbours of user $u \in U$ and denoted by $S(u)$. Then:

$$S(u) = \{v \in U : (uv) \in F, |F(v)| = 1\} = \{v \in U : |F(v)| = 1\} \cap F(v) \quad (1)$$

If user activity level (i.e. the number of publications on the target topic) $u \in U$ is denoted by $r(u)$, then the total activity level of some subset of users $V \subset U$ will be calculated by the formula:

$$R(V) = \sum_{u \in V} r(u) \quad (2)$$

Using formula (2), we get a formula for calculating the weight of a bridge-connected cluster of an arbitrary graph user:

$$C_{WC}(u) = R(S(u)) = \sum_{w \in S(u)} r(w) \quad (3)$$

Weighted contribution centrality is defined as the ratio of the weight of the cluster connected by the bridge to the network core to the total activity level of all clusters in the network. The total activity level of all clusters in the network can be calculated as

$$R = \sum_{u \in V} W(u) \quad (4)$$

Thus, weighted contribution centrality can be expressed as

$$C_{WC}(u) = \frac{W(u)}{R} \quad (5)$$

The code for the Python3 function used to calculate bridges as part of the Python program [23] is shown below:

```python
def calculate_weighted_contribution_centrality(graph, rating):
    centrality = {}
    R = 0 # R = 0 # accumulated weight value of all clusters
    for user, friends in graph.items():
        c = 0 # accumulative value of cluster weight to be connected by user
        # accumulate rating by user's friends
        for friend in friends:
                        # if a user's friend is linked in the column only, add their
                rating
            if len(graph[friend]) == 1:
                c = c + rating[friend]
        centrality[user] = c
        R = R + c
    for user in centrality.keys():
        centrality[user] = centrality[user] / R
    return centrality
```

3.1 Evaluating the Effectiveness of the Bridge Detection Method

To determine the level of influence of bridges from 10 random graphs, the 10 most influential bridges and their associated vertices were removed, as well as those vertices that were isolated after the bridges were removed. Table 1 shows how much the weight of the graphs as a whole and the total weight of the vertices included in the clusters decreased.

As Table 1 indicates, when the 10 most influential bridges are removed from the graphs, the total weight of the graph or the total level of user activity in the graph decreases by an average of 57.8%, indicating a high level of influence of the bridges. At the same time, the total cluster weight decreases by 80.9%, which corresponds to the role played by the 10 most influential bridges in network expansion. From this we can conclude that the network nodes we identified as bridges do contribute significantly to the overall level of network activity. A comparison of the results obtained using centrality by intermediacy and centrality by contribution is presented in (Table 2).

As shown in Table 2, when the 10 most influential bridges along with all their nodes are removed from the graphs, the total weight of the graph decreases approximately equally. This suggests that all three measures are equally effective in revealing the communication abilities of the influential nodes in the network. At the same time, the change in cluster weight is noticeably larger when using weighted contribution centrality.

Table 1. Change in graph weight as a result of removing 10 bridges with their vertices from the graph.

Graph number	Changing graph weight	Changing cluster weights
1	−32,7%	−67,5%
2	−53,0%	−88,6%
3	−44,1%	−65,9%
4	−92,0%	−94,5%
5	−97,3%	−99,7%
6	−72,1%	−92,6%
7	−21,9%	−61,0%
8	−27,2%	−68,3%
9	−48,5%	−72,3%
10	−83,4%	−98,9%
	Average value	*Average value*
	−57,2%	**−80,9%**

Table 2. Change in various centrality measures as a result of removing 10 bridges with their vertices from the graph.

Measure	Changing graph weight	Changing cluster weights
Betweenness centrality	−54,4%	−76,0%
Contribution centrality	−52,7%	−66,4%
Weighted contribution centrality	−57,2%	−80,9%

This is because betweenness centrality and contribution centrality identify the most communicative nodes in the network, including bridges, as opposed to weighted contribution centrality, which only identifies bridges. And since the removal of bridges gives the largest contribution to the reduction in the overall level of network activity, the impact of bridges is greater than that of any other nodes in the network.

Thus, it can be argued that centrality on weighted contribution solves the bridging problem most effectively compared to the other metrics presented.

4 Conclusion

A feature of the weighted contribution centrality measure is that it unambiguously determines whether a node is a bridge in the network configurations described earlier.

Bridges contribute to expanding the size of the network, increasing the number of users involved in the social phenomenon and increasing the overall level of activity of the social network. Blocking the most influential bridges can significantly change the characteristics of the entire network and reduce the overall level of social network activity in a given social phenomenon. Therefore, targeting the most influential bridges is an effective way to reduce social network activity.

The level of informational influence is assessed by ranking the bridges in order of centrality by weighted contribution.

This method of identifying bridges and assessing their informational impact was used as part of an analytical study "political protest propaganda structures in Russia and Belarus", conducted by the SEUSLAB analytical centre LLC. The research included an assessment of the operational significance of the findings and it was presented on the site of the CIS Antiterrorist Centre and the CIS Research Institute for Security Problems in March 2021. Based on the results of the piloting, the Scientific Advisory Board of the CIS Antiterrorist Center drafted an expert opinion on the feasibility of using this method in the information and analytical systems used in the operational and service activities of the Russian Interior Ministry.

References

1. Castells, M.: Networks of Outrage and Hope Social Movements in the Internet Age. Polity, Cambridge (2012)
2. Gerbaudo, P.: Tweets and the Streets. Social Media and Contemporary Activism. Pluto Books, London (2012)
3. Faris, D.M.: Dissent and Revolution in a Digital Age. I.B.Tauris, London (2013). https://doi.org/10.5040/9780755607839
4. Tindall, D.B.: From metaphors to mechanisms: critical issues in networks and social movements research. Soc. Netw. **29**, 160–168 (2007). https://doi.org/10.1016/j.socnet.2006.07.001
5. Bennett, W.L., Segerberg, A.: The logic of connective action. Inf. Commun. Soc. **15**, 739–768 (2012). https://doi.org/10.1080/1369118X.2012.670661
6. Juris, J.S.: Reflections on #Occupy everywhere: social media, public space, and emerging logics of aggregation. Am. Ethnol. **39**, 259–279 (2012). https://doi.org/10.1111/j.1548-1425.2012.01362.x
7. https://www.seuslab.ru/seus
8. https://meduza.io/feature/2018/10/16/politsiya-po-vsey-rossii-pokupaet-sistemy-monito ringa-sotssetey-oni-pomogayut-iskat-ekstremizm-ne-vyhodya-iz-rabochego-kabineta
9. Lü, L., Chen, D., Ren, X.-L., Zhang, Q.-M., Zhang, Y.-C., Zhou, T.: Vital nodes identification in complex networks. Phys. Rep. **650**, 1–63 (2016). https://doi.org/10.1016/j.physrep.2016.06.007
10. Anthonisse, J.M.: The Rush in a Graph. Mathematisch Centrum (mimeo) (1971)
11. Freeman, L.C.: A set of measures of centrality based on betweenness. Sociometry **40**, 35 (1977). https://doi.org/10.2307/3033543
12. Freeman, L.C.: Centrality in social networks conceptual clarification. Soc. Netw. **1**, 215–239 (1978). https://doi.org/10.1016/0378-8733(78)90021-7
13. Opsahl, T., Agneessens, F., Skvoretz, J.: Node centrality in weighted networks: generalizing degree and shortest paths. Soc. Netw. **32**, 245–251 (2010). https://doi.org/10.1016/j.socnet.2010.03.006
14. Kuznetsov, E.N.: Analysis of the structure of network interactions: context-dependent measures of centrality. Management of large systems, pp. 57–82. IPU RAS, Moscow (2019)

15. Wang, H., Hernandez, J.M., van Mieghem, P.: Betweenness centrality in a weighted network. Phys. Rev. E **77**, 046105 (2008). https://doi.org/10.1103/PhysRevE.77.046105
16. van Mieghem, P., van Langen, S.: Influence of the link weight structure on the shortest path. Phys. Rev. E. **71**, 056113 (2005). https://doi.org/10.1103/PhysRevE.71.056113
17. Levandowsky, M., Winter, D.: Distance between sets. Nature **234**, 34–35 (1971). https://doi.org/10.1038/234034a0
18. Wei, H., et al.: Identifying influential nodes based on network representation learning in complex networks. PLOS ONE **13**, e0200091 (2018). https://doi.org/10.1371/journal.pone.0200091
19. Zhang, Q., Karsai, M., Vespignani, A.: Link transmission centrality in large-scale social networks. EPJ Data Sci. **7**(1), 1–16 (2018). https://doi.org/10.1140/epjds/s13688-018-0162-8
20. Ghalmane, Z., El Hassouni, M., Cherifi, C., Cherifi, H.: Centrality in modular networks. EPJ Data Sci. **8**(1), 1–27 (2019). https://doi.org/10.1140/epjds/s13688-019-0195-7
21. Jensen, P., et al.: Detecting global bridges in networks. IMA J. Complex Netw. **4**, 319–329 (2015)
22. Alvarez-Socorro, A.J., Herrera-Almarza, G.C., González-Díaz, L.A.: Eigencentrality based on dissimilarity measures reveals central nodes in complex networks. Sci. Rep. **5**, 17095 (2015). https://doi.org/10.1038/srep17095
23. Rabchevskiy, A.N., Zayakin, V.S.: The program for calculating bridges in cluster networks. Certificate of state registration of computer programs No. 2021616086 of 16 April 2021 (2021)

Theoretical Machine Learning
and Optimization

On the Pareto-Optimal Solutions in the Multimodal Clustering Problem

Mikhail Bogatyrev[✉][ID], Dmitry Orlov, and Tatyana Shestaka

Tula State University, 92 Lenin Avenue, Tula, Russia
okkambo@mail.ru

Abstract. The paper considers the application of multi-objective optimization methods in the problem of multimodal clustering. An overview of the algorithms that deliver Pareto-optimal solutions is made. The use of multi-objective optimization makes it possible to interpret clusters in terms of several criteria. An evolutionary multi-objective algorithm for multimodal clustering is proposed. The algorithm was tested on two and three-dimensional formal contexts that simulate data on myocardial infarction complications in patients of various ages and medical records. Clustering is used to extract facts as specific combinations of data from three sets. The simulation results were compared with the known methods of Formal Concept Analysis.

Keywords: Multi-objective optimization · Multimodal clustering · Pareto optimization · Fact extraction

1 Introduction

Classical cluster analysis is based on dividing a single set of objects into disjoint subsets being clusters. At the same time, despite the wide variety of proximity measures of objects used here, the problem of interpreting the obtained clusters remains urgent in cluster analysis. The clusters interpretation is always performed in the context of the applied proximity measure, and for heterogeneous data from different domains, such an interpretation based on a single proximity measure may not be correct.

Multimodal clustering involves not one, but several sets of objects to be clustered simultaneously. Such sets can be formed by heterogeneous data. A multimodal cluster is a subset in the form of combinations of objects from different sets. The very fact of the combination of certain objects in a multimodal cluster can carry important information and serve as the basis for clusters interpretation.

Formal Concept Analysis (FCA) [3] offers its own approach to multimodal clustering. In FCA, clustering is used not on one, but on two, three and, in general, on an arbitrary number of sets, this is biclustering, triclustering and multimodal clustering.

E. Burnaev et al. (Eds.): AIST 2021, CCIS 1573, pp. 179–194, 2022.
https://doi.org/10.1007/978-3-031-15168-2_15

When clustering is performed on multiple sets, there may be multiple measures of the proximity of objects from those sets. This is especially true for heterogeneous data presented in such sets, which makes the use of multi-objective optimization in demand.

In this paper, we propose two innovations to FCA technology of multimodal clustering. The first innovation is the use of cluster parameters applied in FCA as independent contradictory optimization criteria. The second innovation is an evolutionary algorithm of multi-criteria optimization.

The rest of the paper is organized as follows.

In Sect. 2, there is the detailed motivation of this work with the analysis of related works. In Sect. 3, the proposed algorithm for multimodal clustering is presented. The results of the experimental study of the proposed approach are presented in Sect. 4. They are illustrated on the task of phenotyping of disease of myocardial infarction. Results of the experimental study are discussed for further implementations in Sect. 5.

2 Motivation and Related Work

We note a few, in our opinion, important provisions that determine the motivation of this work.

2.1 Multimodal Clustering Problem

Multimodal clustering is the clustering of multimodal objects. Under modality, it is often assumed one of more attributes which describe a specific characteristic of an object. When an object has multiple modalities describing it, it can be called a multimodal object [1].

Multimodal clustering involves not one, but several sets of data to be clustered simultaneously. Such sets can be formed by heterogeneous data which have high variability of data types and formats [2]. Therefore, the terms *multimodal* and *heterogeneous* can characterize the same data.

Clustering algorithms for a single set of objects use various measures of the proximity of data instances, which are usually determined on pairs of elements of the clustered set. In multimodal clustering, in addition to traditional proximity measures, characteristics describing the relationships between multimodal sets can be used.

If such characteristic is formal relation $R \subseteq D_1 \times D_2 \times \ldots \times D_n$ on data domains $D_1, D_2, \ldots, D_n$ then we have so called *formal context* as an $n+1$ set:

$$\mathbb{K} = < K_1, K_2, \ldots, K_n, R > \tag{1}$$

where $K_i \subseteq D_i$. *Multimodal clusters* on the context (1) are n - sets

$$\mathbb{C} = < X_1, X_2, \ldots, X_n > \tag{2}$$

which have the closure property [4]:

$$\forall u = (x_1, x_2, \ldots, x_n) \in X_1, X_2, \ldots, X_n, u \in R, \tag{3}$$

$\forall j = 1, 2, \ldots, n, \forall x_j \in D_j \backslash X_j < X_1, \ldots, X_j \cup \{x_j\}, \ldots, X_n >$ does not satisfy (3).

A multimodal cluster is a subset in the form of combinations of elements from different sets K_i. It is also defined as a closed n-set [4] since the closure property (3) provides its "self-sufficiency": it cannot be enlarged without violating (2).

The definition of a multimodal cluster (2)–(3) does not use a proximity measure. Domain elements are linked to each other by means of the relation R and these links define combinations of domain elements in the form of multimodal clusters.

Multimodal clustering on the contexts constructed by using relations is studied in the Formal Concept Analysis (FCA). Classical FCA uses formal contexts with binary relation R and biclustering in the form of formal concepts, which make up a *conceptual lattice* [3].

In the FCA, triclustering is a generalization of biclustering [5]. However, the appearance of the third set in the data presentation fundamentally changes the situation, and triclustering algorithms are not built by simple scaling of biclustering ones. An overview of triclustering algorithms can be found in [6]. The generalization of the approach based on the construction of concept lattices to the n-dimensional case of multimodal clustering is presented in [7].

Formal concepts on multimodal formal context are those multimodal clusters where *for all* $u = (x_1, x_2, \ldots, x_k) \in X_1, X_2, \ldots, X_k, u \in R$ and k is maximally possible. In other words, they are the largest possible k-dimensional hypercubes completely filled with units. The concept of the density of a multimodal cluster is introduced in the FCA and formal concepts are interpreted as absolutely dense clusters.

There are some practical arguments in favor of studying multimodal clusters as none dense concepts. In this case, in addition to the density of clusters, their other characteristics are introduced: *volume, modality, diversity*, as well as *coverage* of the context [8]. These characteristics illustrate the quality of multimodal clustering and in some cases help to interpret the contents of clusters.

Having a set of clustering quality parameters, the multimodal clustering problem is formulated as an optimization problem in which the extremum of the criterion based on these parameters is searched for [8,9]. In fact, some of the mentioned parameters, for example, the volume of clusters and their density, form conflicting criteria.

One motivation of our work is that multimodal clustering may be formulated as a multi-objective optimization problem. Therefore, our task is to propose a solution to the problem of multimodal clustering as a multi-objective optimization problem.

2.2   Evolutionary Approach in Multi-objective Optimization

The evolutionary approach to multi-objective optimization is based on Evolutionary computation [11]. Evolutionary computation is a collection of algorithms for solving global optimization problems that use the evolution of solutions. The first known evolutionary algorithm is the genetic algorithm [11], which realizes a probabilistic optimization method based on the biological principles of evolution. Now there are more than 40 different types of bio-inspired optimization algorithms [12].

The typical evolutionary algorithm uses the terminology of genetic algorithms and includes the following steps.

---

1 $t := 0$;
2 *Compute initial population P_0;*
3 **while** *stopping condition not fulfilled* **do**
4   *select individuals for reproduction;*
5   *create offsprings by crossing individuals;*
6   *mutate some individuals;*
7   *compute new generation;*
8     $t := t + 1$;

Evolutionary algorithms manipulate encoded versions of the problem solutions instead of the solutions themselves. Every solution to a problem is represented as a chromosome. The chromosome is the string of certain length containing *genes* being 0/1 values (binary encoding) or natural numbers. The use of encoding with chromosomes is intended by the desire to reproduce the biological mechanism of heredity in genetic algorithms. In general, the gene may be a symbol from an *encoding alphabet*. In clustering, there are various encodings [10,13], which differ in how the clusters are encoded. For example, whether a cluster is encoded by one chromosome or one chromosome encodes multiple clusters.

Computing Initial Population. The initial population is a randomly generated set of N chromosomes. At every step t of the evolution, several problem solutions are forming a *population P_t*. A population consists of *individuals* also named *chromosomes*. Evolutionary algorithms use populations with a constant and variable number of chromosomes.

Selection of Individuals for Reproduction. Some *individuals* from the set of chromosomes are selected for creating a new population. Selection works using values of the so-called *fitness function* which characterize the quality of the solution. Therefore, the higher fitness chromosomes have more opportunity to be selected than the lower ones and a good solution is always alive in the next generation. There are several variants of selection such as *proportional selection, roulette wheel selection, tournament selection,* etc. [17].

Computing New Generation. The best individuals selected at each step of evolution become parents of new elements of the population for the next step. Descendant solutions are obtained from parent solutions using mutation and recombination operators, which are random in nature. *Crossing individuals* is made by crossover, the genetic operator that mixes two chromosomes to form a new offspring. It does mixing by replacing fragments of chromosome code divided in certain one or several randomly selected points. *Mutation in some individuals* involves the modification of gene values by randomly selecting a new value from the gene alphabet at a random point in the strings of genes.

Evolution of Solutions. Evolution starts by applying evolutionary operators A to population P_0 and further iteratively so that for every $P_{k+1} = \mathbb{A}(P_k)$ exists at least one

$$f[\varphi(p_{k+1})] \geq f[\varphi(p_k)], \tag{4}$$

where $p_k \in P_k$ and $p_{k+1} \in P_{k+1}, \varphi$ is a mapping $\varphi : P_i \to \mathbb{S}$, where $\mathbb{S} = \{S_i\}$ is the set of solutions, f is the fitness function defined on solutions. Evolution finishes in accordance with the stopping criterion. Most often, the criterion for stopping is the immutability of the fitness function values over several steps of evolution.

The evolution of solutions organized in this way leads the population of chromosomes to a state corresponding to the local or global extremum of the fitness function. The fitness function is calculated for each chromosome and, as a rule, generalized to the whole population. Several optimal solutions may remain at the final step of evolution.

These features of Evolutionary computation characterize them as a very suitable tool for solving multi-objective optimization problems with conflicting criteria, where no single solution exists that simultaneously optimizes each objective. Therefore, the evolutionary approach is known in clustering [10].

Pareto-optimal Solutions. The concept of Pareto optimality belongs to the main stream in the domain of multi-objective optimization. There is a significant number of multi-objective evolutionary algorithms (MOEAs) proposed mainly in two decades from the 1990s until the 2010s. Detailed comparisons of different MOEAs for clustering can be found in [10,14,15]. Among them there are MOEAs focused on obtaining Pareto optimal solutions: Niched Pareto Genetic Algorithm (NPGA), Strength Pareto Evolutionary Algorithm (SPEA), Non-dominated Sorting Genetic Algorithm (NSGA), and others reviewed in [15].

Pareto optimality from the viewpoint of maximization optimization problem may be given as follows. Let $\mathbb{S} = \{S_i\}$ is the set of solutions of multi-objective optimization problem, $i = 1, 2, \ldots, n, F = \{f_i\}, j = 1, 2, \ldots, m$ is the set of objective functions. Every solution is characterized by vector $\mathbf{f}_i = \{f_i(S_i)\}$. One feasible solution $\mathbf{f}_i$ is said to *dominate* another feasible solution $\mathbf{f}_i$ if and only if $f_j(S_i) \geq f_j(S_k)$ *for all* $j = 1, 2, \ldots, m$ and $f_j(S_i) \geq f_j(S_k)$ for least one objective function $f_d, d \in \{1, 2, \ldots, m\}$. A solution is said to be *Pareto optimal* if it is not dominated by any other solution. A Pareto optimal solution cannot be improved with respect to any objective function without worsening value at least one other objective function. The set of all feasible non-dominated solutions is referred to as the Pareto optimal set, and for a given *Pareto optimal set*, the corresponding objective function values in the objective space are called the *Pareto front* [16].

Evolutionary algorithms have certain advantages in implementing Pareto-optimal solutions to multi-objective optimization problems. Among them, there is one important which consists in the fact that the presence of a population of solutions supported by the algorithm allows you to naturally organize the formation of the Pareto front.

The considered features of evolutionary algorithms determine our motivation for their application in multimodal clustering.

3   Evolutionary Multi-objective Algorithm for Multimodal Clustering

The problem of multimodal clustering on formal contexts has the features discussed above. We consider these features in the evolutionary multimodal clustering algorithm.

As a proximity measure in the clustering problem, we do not consider the distances between the objects being clustered, but compare the clusters by their characteristics. We study these characteristics in pairs and solve the multi-objective optimization problem for two contradictory criteria. The following characteristics of multimodal clusters are used in the clustering algorithm.

Cluster Density and Volume. For a cluster (2) its density is defined as

$$d(\mathbb{C}) = \frac{|R \cap (X_1 \times X_2 \times \ldots \times X_n)|}{|X_1| \times |X_2| \times \ldots \times |X_n|} \tag{5}$$

and volume of a cluster has the following form

$$v(\mathbb{C}) = |X_1| \times |X_2| \times \ldots \times |X_n| \tag{6}$$

Cluster density and volume are contradictory criteria for cluster quality. A large and dense cluster is interesting because combinations of elements of its subsets set a property that manifests itself on a large number of elements and, possibly, means a regularity. However, often the clustered data is sparse and the existence of large and dense clusters on them is unlikely. Therefore, when selecting clusters, a trade-off between density and volume is necessary.

Coverage and Diversity. These two cluster characteristics were discussed and defined for the triclustering problem in [8]. They may be generalized for multimodal clustering. Coverage is defined as a fraction of the tuples of the context included in at least one of the multimodal clusters. This can be defined by analogy with the definition in [8]:

$$\sigma(\Omega) = \sum_{(x_1, x_2, \ldots, x_n) \in R} [(x_1, x_2, \ldots, x_n) \in \bigcup_{(x_1, x_2, \ldots, x_n) \in \Omega} (x_1 \times x_2 \times \ldots \times x_n)]/|R|, \tag{7}$$

where Ω is a set of multimodal clusters.

The data of the sets that make up the cluster modalities have different meanings. Sometimes it is important to control the coverage of the context by some subset of the cluster, for example, as will be shown below, by a subset of patients.

In this case, in the expression (7), instead of a whole tuple $(x_1, x_2, \ldots, x_n)$ it is used one of its elements.

The definition of cluster diversity given in [8] is valid for multimodal clusters:

$$\tau(\Omega) = 1 - \frac{\sum_j \sum_{i<j} \gamma(\Omega_i, \Omega_j)}{\dfrac{|\Omega|(|\Omega| - 1)}{2}}, \tag{8}$$

where $\gamma(\Omega_i, \Omega_j)$ is an intersection function which is equal to 1 if clusters Ω_i, Ω_j intersect at least one of their subsets and 0 otherwise.

Encoding Scheme. After analyzing the several variants of chromosome encoding [14,15], we settled on the binary scheme organized according to the principle of "one chromosome - one cluster". If the formal context has modality n then a chromosome has n sections. In the sections, a number of a gene is the number of an element of a corresponding set in a multimodal context. The units in the chromosome correspond to the numbers of elements of subsets of the context that fall into this cluster. This scheme is illustrated in Fig. 1.

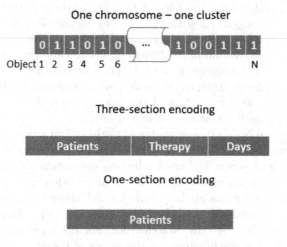

Fig. 1. The chromosome encoding scheme "one chromosome - one cluster".

On the top of Fig. 1 there is a chromosome having N positions corresponding to N objects to be clustered. This chromosome represents a cluster containing those objects that have units in their positions in the chromosome.

The picture in the middle of Fig. 1 illustrates three-section chromosomes that we used in clustering three-dimensional formal contexts below. Crossover and mutation may operate in all sections separately to maximize search space coverage.

Also there is encoding with one-section chromosome for formal context with modality n which is shown at the bottom of Fig. 1. With this encoding, the algorithm works not only on chromosomes, but also in a formal context, providing synchronous processing of data from all n sets.

This binary encoding scheme is not compact because for large contexts with high modality the chromosomes will be very long. Nevertheless, in the task of clustering, it is much more convenient to work with such chromosomes than with chromosomes with a more compact length. Explicit representation of clusters in the form of separate chromosomes does not require additional computations, which are necessary for other encodings. In addition, handling large binary strings is not a problem.

However, among the new chromosomes generated in this way, there may be incorrect chromosomes, which do not correspond to the data in the original context. A special function is needed to filter out the wrong chromosomes.

3.1   Elitist Nondominated Sorting

A Pareto front is a subset of a population of chromosomes corresponding to nondominated solutions. When forming Pareto fronts by evolutionary algorithm, they are being assembled sequentially as new chromosomes are generated during the evolution of solutions. Evolution is performed by applying genetic operators of selection, mutation and crossover. If the probability of mutation and crossover is high enough and the crossover is not tied to the peculiarities of chromosome encoding, then the algorithm performs random uncontrolled walks in the search space. This guarantees that the algorithm will explore most of the search space to find the global extremum of the fitness function. However, such walks reduce the convergence of the algorithm and, in principle, do not exclude its cycling in the regions of local extrema. Moreover, when calculating the Pareto front, random walks can lead to a "loss of the front", when the constructed Pareto front is destroyed at the next step of evolution.

A well-known solution that contributes to the construction of controlled evolution is *elitism* [15,18]. Elitism may be considered as an operator which preserves the better of parent and child solutions (or populations or Pareto fronts) so that a previously found better solution is never deleted. At each step of evolution, the better solutions in the population are those chromosomes that have maximum or minimum values of the fitness function, depending on whether its maximum or minimum is being searched for. For the selection of such chromosomes, a threshold of the difference in the values of the fitness function is introduced: $\forall i,j : |f(c_i^{(k)}) - f(c_j^{(k)})| \leq h$, where $c_i^{(k)}, c_j^{(k)}$ are chromosomes created at the k-th step of evolution.

In the case of Pareto optimization, elitism is associated with dominance, and it is necessary to preserve not individual solutions, but, if possible, the entire front. In the MOEAs, elitism is realized as *nondominated sorting* [18].

3.2   Algorithm and Its Functions

Our algorithm is based on the NSGA-II algorithm [18] which is one of the popularly known MOEAs. It produces non-dominated solutions of multi-objective optimization, the diversity is maintained in the solution, and the elitist principle has been used in the algorithm.

We have made several changes to the NSGA-II. It was adapted for clustering. We also added functions to the algorithm for visualizing Pareto fronts.

The algorithm is shown below. The purpose of most of the functions used in it is obvious from their names and from the comments in the algorithm. Some of the functions used in the algorithm have several variants.

doSelection function realizes selection chromosomes according to the selection method. There are *proportional, random universal, tournament* and *truncation* selection methods [14] realized in the algorithm. The specific selection method is picked through the user interface.

Algorithm 1: Evolutionary multi-objective clustering algorithm

Input: tensor is multidimensional context as the set of n samples on the axes of measurements

Parameters:

sizePop is the size of population of chromosomes;
numpoints is the number of points of crossover;
mutationRate is the probability of mutation;
crossoverRate is the probability of crossover;
limitPop is the maximal number of populations;
countPop is the number of steps of evolution;
popFitness is the value of the fitness function for the entire population.
historyPop is stores all the populations

Output: clusters is the set of clusters in the form of a set of subsets.

1   *population ← createPopulation[tensor, sizePop]* creating a population of chromosomes *chrom*

2   **while** *countPop ≤ limitPop* **do**

3   **for** *all chrom* **do**

4   *clusterDensity[chrom, tensor]*

5   *clusterVolume[chrom, tensor]*

6   *fitnessFunction[chrom, tensor, coefDensity, coefSize]*

7   *doSelection[chrom, popFitness]*

8   *doMultipleCrossover[chrom1, chrom2, numpoints, tensor]*

9   *doMutation[chrom, mutationRate, tensor]*

10   *popFitness[population]* calculating the value of the fitness function for the entire population.

11   *combPop ← historyPop ∪ population* provides the elitism of the best chromosomes

12   *{front, rest} ← survivorSelection[combPop, popSize]* obtaining front and rear chromosomes

13   *historyPop ← historyPop ∪ front* replenishment with front-end chromosomes

14   *visualizePop[front, rest]* visualization on the Pareto front

15   **for** *all chrom* **do**

16   *chrom ← getSubTensorChrom[chrom, tensor]* forming clusters from tensor

---

The *doMultipleCrossover* function, in addition to performing a crossover, accesses the original tensor in order to filter out the wrong chromosomes. We have also provided the crossover mode which is performed only in certain sections of chromosomes.

4   Experiments

An experimental study of the proposed approach was carried out on the two data sets. The first set is the Myocardial Infarction Complications Data Set (MICDS) [19]. It contains information about 1700 patients having the disease of myocardial infarction and its complications. The whole formal context has 1700 objects and 123 attributes. Among attributes, there are ones about patients (ID only), their anamnesis, their treatment methods, and complications after treatment. An attribute may be binary or has a value as a natural number.

The second data set is a triadic formal context containing data about offenses committed by juveniles [20]. Its objects are the offense names, the attributes are the age group which had the most amount of certain sort offense, the conditions are the years the offenses take place.

Our first goal was to test the performance of the algorithm in the building sets of non-dominated clustering solutions with acceptable performance. This refers to the processing of long chromosomes of about 2000 genes per line.

The second goal was to study the results of clustering as sources of facts. We were primarily interested not in the number of clusters of different densities and volumes, but in their content, which could be interpreted as facts. To do this, we selected certain formal contexts from the whole MICDS, which relate to the problem of phenotyping diseases. Disease phenotyping refers to the determination of the form of the disease based on the clinical profile. A clinical profile is a cluster that can include various data describing both the disease itself and the methods of its treatment, as well as the conditions of patients and sometimes the treatment results.

4.1   Testing the Algorithm

At first, we investigated the convergence of the algorithm and the construction of Pareto fronts. Even on the small population of 10 chromosomes, the algorithm quickly builds a Pareto front as it is shown in Fig. 2.

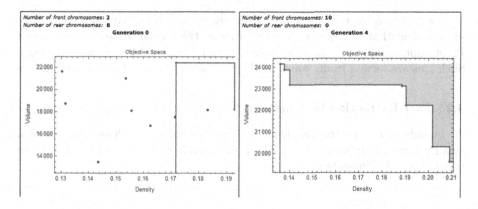

Fig. 2. The initial and final Pareto front in the two-objective clustering.

In the experiment illustrated in Fig. 2, the Pareto front on a formal context of 1700 objects and 43 attributes was constructed in 4 steps. The final population, forming the Pareto front, contains 7 different chromosomes being clusters. Fast convergence is a known feature of genetic algorithms when they find solutions corresponding to the local extrema of the fitness function. It usually takes place when mutation probability has small values. In the experiment on Fig. 2 mutation probability was 0.01, there was a two-point crossover with probability = 0.8 and proportional selection. Therefore, the Pareto front in Fig. 2 can be called local. It contains large and not dense clusters which are not informative.

Coverage as Characteristic of Globality. We need to expand the search space to find clusters being as dense as possible. A possible but not guaranteed to be successful way to achieve this is to increase the probability of mutations and increase the size of the population [17]. For Pareto-optimal optimization, this problem is formulated as the problem of searching the global Pareto front. We apply calculations of cluster coverage (7) as a characteristic of the coverage of the search space. There are general and individual coverages being calculated for the whole population and for particular clusters. In Fig. 3 the dependence of the coverage value on the population size is illustrated.

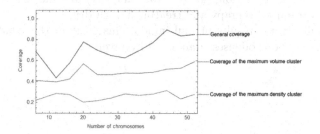

Fig. 3. Dependence of the coverage on the population size.

For each specific context, there is a minimum number of chromosomes (population size) that provides a given coverage of the formal context. We consider Pareto-optimal clusters as ones having maximal volume, density and coverage. Such clusters form a front, which in this case serves as a global Pareto front.

4.2   Fact Extraction by Clustering

As already noted, in the problem of disease phenotyping, dense clusters of a small volume are informative. In Fig. 4 an example of Pareto front and a dense small cluster is illustrated.

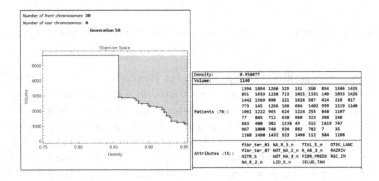

Fig. 4. An example of a Pareto front and a dense small cluster (at the bottom right of the front) that has attributes related to the treatment of a heart attack and further complications.

The cluster in Fig. 4 includes patients who are characterized by a combination of attributes related to the applied therapy and the results of treatment. All patients received variants of fibrinolytic therapy and treatment with conventional drugs for a heart attack. Among complications, there are atrial fibrillation, ventricular tachycardia, pulmonary edema, myocardial rupture, and recurrence of myocardial infarction.

Therefore, this cluster contains the fact of the presence of serious complications of myocardial infarction, possibly (density $\neq 1$) in some of the 76 patients who received standard therapy and treatment with drugs.

Figure 5 illustrates results of multimodal clustering of the triadic data set containing data about offenses committed by juveniles.

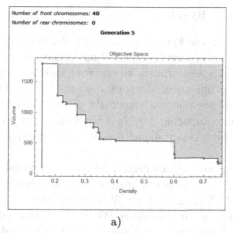

a)

Density:	0.75			
Volume:	180			
Offense names (15):	Burglary Drug Abuse Violations Drunkenness Drive Under the Influence	Embezzlement Forgery and Counterfeiting Gambling Murder and Nonnegligent Homicide	Liquor Laws Manslaughter by Negligence All Other Offenses Runaway	Sex Offenses Suspicion Vandalism
Age/group (1):	m_17			
Years (12):	1994 1999 2004 2009 1996 2000 2005 2015 1997 2002 2007 2016			
Individual coverage:	0.0372671			

b)

Density:	0.207031				
Volume:	1280				
Offense names (20):	Aggravated Assault Simple Assault Burglary Curfew and Loitering Law Violations	Drug Abuse Violations Drunkenness Drive Under the Influence Embezzlement	Offenses Against the Family and Children Fraud Gambling Murder and Nonnegligent Homicide	Liquor Laws Larceny Prostitution and Commercialized Vice Robbery	Rape Runaway Stolen Property Weapons
Age group (4):	m_17 f_13_14 m_16 f_16				
Years (16):	1995 2003 2007 2011 1998 2004 2008 2012 2001 2005 2009 2014 2002 2006 2010 2016				
Individual coverage:	0.26501				

c)

Fig. 5. An example of a Pareto front and two clusters illustrate the involvement of adolescents in offenses.

The Pareto front is shown in Fig. 5 a). It was built on the 5 th generation with mutation probability 0.01, two-point crossover with probability = 0.8, and proportional selection. The Pareto front has 16 clusters, two of which are shown in Fig. 5 b) and Fig. 5 c). Cluster on Fig. 5 b) informs about offenses typical for 17-year-old boys. Cluster on Fig. 5 c) shows such offenses in which 13- and 14-year-old girls were involved together with older teenagers. The analysis of these facts of involvement in offenses of adolescents of different ages in these two and other clusters allows us to determine the nature of offenses, their prevalence by year, and by age groups.

4.3   Comparison of Results

The NSGA-II algorithm is known as the most successful algorithm for finding Pareto-optimal solutions to multi-objective optimization problems. Comparisons of the operation of this algorithm with the well-known evolutionary algorithms of multi-objective optimization are contained in [15,18]. The data and fitness functions that we use are not fundamentally different from the data and fitness functions used when comparing NSGA-II with other algorithms. Therefore, we did not use other evolutionary algorithms of multi-criteria optimization in our study.

We performed Pareto-optimal clustering on two-dimensional formal contexts in order to be able to compare our results with the results of biclustering performed by standard FCA algorithms [3]. Biclustering with FCA allows one to build association rules and implications. We have constructed conceptual lattices for our contexts and association rules on them. Conceptual lattice constructed on the context presented in the experiments contains 66910 formal concepts.

Next, we checked whether there are semantic coincidences between association rules and cluster interpretations.

For all found informative clusters - facts, we have found the corresponding association rules on the attributes in the conceptual lattice.

5   Discussion and Conclusion

This paper proposes an approach to multimodal clustering on multidimensional formal contexts using evolutionary computation. Here we present the results of clustering on two-dimensional formal contexts, but the algorithm also works on multimodal contexts of any dimension.

The presented experimental results reflect the initial stage of research in this area.

Only one encoding scheme, which is most convenient for controlling the algorithm, is studied in detail. In this scheme, the maximum number of clusters is set, which coincides with the size of the population. Other known schemes [10,13] do not limit the initial number of clusters, but, according to our estimates, reduce the performance of the algorithm due to many additional computations.

In the future, it is planned to do the following.

1. Create a tool for a detailed study of clusters belonging to the same Pareto front.
2. Evaluate the informativeness of the obtained clusters not manually, but using a user interface focused on end-user.
3. To make a comparison with other coding schemes of evolutionary algorithms.

We hope that the proposed approach to multimodal clustering complements the existing FCA methods.

Acknowledgments.. The reported study was funded by the Russian Foundation of Basic Research, the research projects 19-07-01178, 20-07-00055, and RFBR and Tula Region research project 19-47-710007. Our special thanks are to Robin Gruna for the visualization of Pareto fronts and for the second anonymous reviewer for detailed and useful analyses of this work.

References

1. Lahat, D., Adalı, D., Jutten, C.: Multimodal data fusion: an overview of methods, challenges and prospects. Proc. IEEE Inst. Electric. Electron. Eng. Multim. Data Fusion **103**(9), 1449–1477 (2015)
2. Wang, L.: Heterogeneous data and big data analytics. Autom. Control Inf. Sci. **3**(1), 8–15 (2017). https://doi.org/10.12691/acis-3-1-3
3. Ganter, B., Stumme, G., Wille, R. (eds.): Formal concept analysis. LNCS (LNAI), vol. 3626. Springer, Heidelberg (2005). https://doi.org/10.1007/978-3-540-31881-1
4. Cerf, L., Besson, J., Robardet, C., Boulicaut, J.F.: Closed patterns meet N-ary relations. ACM Trans. Knowl. Discov. Data. **3**(1), 36 (2009)
5. Ignatov, D.I., Kuznetsov, S.O., Zhukov, L.E., Poelmans, J.: Can triconcepts become triclusters? Int. J. Gen. Syst. **42**, 6 (2013)
6. Gnatyshak, D.V., Ignatov, D.I., Kuznetsov, S.O.: From triadic FCA to triclustering: experimental comparison of some triclustering algorithms. In: Proceedings of the Tenth International Conference on Concept Lattices and Their Applications (CLA 2013), pp. 249–260. Laboratory L3i, University of La Rochelle, La Rochelle (2013)
7. Voutsadakis, G.: Polyadic concept analysis. Order **19**(3), 295–304 (2002)
8. Ignatov, D.I., Gnatyshak, D.V., Kuznetsov, S.O., Mirkin, B.G.: Triadic formal concept analysis and triclustering: searching for optimal patterns. Mach. Learn. 101, 271–302 (2015) https://doi.org/10.1007/s10994-015-5487-y
9. Mirkin, B.G., Kramarenko, A.V.: Approximate bicluster and tricluster boxes in the analysis of binary data. In: Kuznetsov, S.O., Ślęzak, D., Hepting, D.H., Mirkin, B.G. (eds.) RSFDGrC 2011. LNCS (LNAI), vol. 6743, pp. 248–256. Springer, Heidelberg (2011). https://doi.org/10.1007/978-3-642-21881-1_40
10. Hruschka, E., Campello, R., Freitas, A., de Carballo, A.: A survey of evolutionary algorithms for clustering. IEEE Trans. Evolution. Comput. **39**, 133–155 (2009). https://doi.org/10.1109/TSMCC.2008.2007252
11. Simon, D.: Evolutionary optimization algorithms. In: Biologically-Inspired and Population-Based Approaches to Computer Intelligence, 1st edn. Wiley, Hoboken (2013)
12. Valdez, F., Castillo, O., Melin, P.: Bio-inspired algorithms and its applications for optimization in fuzzy clustering. Algorithms **14**, 122 (2021)
13. Cole, R.M.: Clustering with genetic algorithms, MSc Thesis, University of Western Australia, Australia (1998)
14. Deb, K.: Multi-objective Optimization Using Evolutionary Algorithms. Wiley, London (2001)
15. Mukhopadhyay, A., Maulik, U., Bandyopadhyay, S., Coello, C.: Survey of multiobjective evolutionary algorithms for data mining: part II. IEEE Trans. Evolution. Comput. **18**(1), 20–35 (2014)
16. Abdullah Konaka, A., Coitb, D., Smith, A.: Multi-objective optimization using genetic algorithms: a tutorial. Reliab. Eng. Syst. Safety **91**, 992–1007 (2006)

17. Goldberg, D.E.: Genetic Algorithms in Search Optimization and Machine Learning. Addison-Wesley, Reading (1989)
18. Deb, A.P., Agrawal, S., Meyarivan, T.: A fast and elitist multiobjective genetic algorithm: NSGA-II. IEEE Trans. Evolution. Comput. **6**, 182–197 (2002)
19. Myocardial Infarction Complications Data Set. http://archive.ics.uci.edu/ml/machine-learning-databases/00579/
20. Juvenile Triadic Data Set. http://fca-tools-bundle.com/view-context/60b92250 ef71886d7336e49e

On Several Edge-Disjoint MSTs
with Given Diameter in Undirected
Graph with Exponentially Distributed
Edge Weights

Eduard Kh. Gimadi[1,2] , Aleksandr S. Shevyakov[2(✉)],
and Aleksandr A. Shtepa[2]

[1] Sobolev Institute of Mathematics, Prosp. Akad. Koptyuga, 4, Novosibirsk, Russia
gimadi@math.nsc.ru
[2] Novosibirsk State University, Novosibirsk, Russia
shevash.97@gmail.com
http://www.math.nsc.ru/

Abstract. In this work we consider the m-d-UMST problem. The m-d-UMST is to find m edge-disjoint spanning trees with the given diameter d of a minimal weight in given n-vertex weighted Undirected graph $G = (V, E)$. We give an $O(n^2)$-time approximation algorithm solving this problem. Then we prove its asymptotic optimality in the case of complete undirected graphs. We also assume that weights of graph edges are independent random variables identically distributed from the class of biased exponential distribution.

Keywords: Given-diameter minimal spanning tree · Probabilistic analysis · Asymptotic optimality · Biased exponential distribution

1   Introduction

Let us consider the following problem: you have n hubs that can accept and send information. You know the distance between each pair of hubs and want to lay a communication channel such that any 2 hubs could transfer information to each other. Your goal is to use the channel of the minimum total length.

Mathematically this problem is described as the Minimum Spanning Tree (MST) problem. It is an one of the classic discrete optimization problems. Given undirected weighted graph $G = (V, E)$, MST is to find a spanning tree of a minimal weight. This problem was well explored in the middle of the last century. There are several polynomial algorithms solving this problem. The most famous of them were proposed by Boruvka (1926), Kruskal (1956) and Prim (1957).

Supported by the program of fundamental scientific researches of the SB RAS, project No. 0314-2019-0014, and by the Russian Foundation for Basic Research, project No. 20-31-90091.

E. Burnaev et al. (Eds.): AIST 2021, CCIS 1573, pp. 195–206, 2022.
https://doi.org/10.1007/978-3-031-15168-2_16

These algorithms have complexities $O(m \log n)$, $O(m \log m)$ and $O(n^2)$, where $m = |E|$ and $n = |V|$.

Let us complicate the initial problem a little bit. What if you want to build a channel such that the signal coming from the hub A to the hub B don't visit a big numbers of other hubs. It could be vital if the signal changes slightly going through each channel. It's an example of generalization MST problem that is called diameter-bounded MST problem. The diameter of a tree is the number of edges in the longest simple path within the tree connecting a pair of vertices. For this problem, given a graph and a number $d = d_n$, the goal is to find in the graph a spanning tree T_n of minimal total weight having its diameter bounded above to given number d, or from below to given number d. Both problems are NP-hard in general.

The bounded above MST problem is polynomially solvable for diameters two or three, and NP-hard for any diameter between 4 and $(n-1)$, even for the edge weights equal to 1 or 2 [4, p. 206]. The MST problem bounded from below is NP-hard, because its particular case for $d = n - 1$ is the problem "Hamiltonian Path" [4].

Recently, we decided to modify this problem. We started to consider the diameter of the desired spanning tree as a given number. In the work [6] we gave a probabilistic analysis of an effective algorithm for solving a given-diameter MST problem in the case of directed complete graph. Unfortunately, the algorithm analysis, presented in this work becomes unacceptable for a problem on undirected graphs. The appearance of the difficulty of probabilistic analysis in the case of the undirected graph arises from the need to take into account the possible dependence between different objects (random variables) in the course of the algorithm.

Now we are interested in the generalization of the problem. Let us return to the example with the information channel. Sometimes you need to build several disjoint channels to improve reliability of the system or to be able to transfer several signals at once. Mathematically it means that you need to construct m edge-disjoint spanning trees with the diameter equaled d of minimum total weight.

In our previous work [7] we introduced an asymptotically optimal approach to solve a given-diameter several edge-disjoint MST problem in the Undirected complete graph (m-d-UMST) under conditions that weights of graph edges are independent and identically distributed random variables with uniform distribution.

In current paper we prove that the same algorithm solves a m-d-UMST problem in case of weights of graph edges are independent and identically distributed random variables from the class of biased exponential distribution. We also provide conditions for this algorithm to be asymptotically optimal.

2   Finding Several Edge-Disjoined MSTs with a Given Diameter

Given a complete weighted n-vertex graph $G = (V, E)$ and positive digits m, d such that $m(d + 1) \leq n$, the problem is to find m edge-disjoint spanning trees

$T_1, \ldots, T_m$ with a given diameter $d = d_n \leq n/m$, such that minimize their total weight.

Description of the Algorithm $\mathcal{A}$

Preliminary Step 0. Choose in the graph G an arbitrary $(n - m(d+1))$-vertex subset V', and remaining $m(d + 1)$ vertices split arbitrary on $(d + 1)$-vertex subsets $\{V_1, \ldots, V_m\}$.

Step 1. For each $i = 1, \ldots, m$, starting at arbitrary vertex in the subgraph $G(V_i)$, construct in it a Hamiltonian path P_i of a length $d = d_n$, using the approach "Go to the nearest unvisited vertex".

Step 2. For each pair of paths P_i and P_j, $1 \leq i < j, \leq m$, we connect them in a special way by the set E_{ij} of $2(d + 1)$ disjoint edges, so that the constructed subgraph was composed of two $2(d + 1)$-vertex edge-disjoint subtrees with a diameter equals d. We represent each path as two halves (first and second). We construct the set of connecting edges as follows.

We assume without loss of generality that d is odd.

2.1. Connect each inner vertex of the first half of the path P_i by the shortest edge to the inner vertex of the first half of the path P_j.

2.2. Connect each inner vertex of the second half of the path P_i by the shortest edge to the inner vertex of the second half of the path P_j.

2.3. Connect each inner vertex of the first half of the path P_j by the shortest edge to the inner vertex of the second half of the path P_i.

2.4. Connect each inner vertex of the second half of the path P_j by the shortest edge to the inner vertex of the first half of the path P_i.

2.5. Connect each end vertex of the path P_i by the shortest edge to the inner vertex of the path P_j.

2.6. Connect each end vertex of the path P_j by the shortest edge to the inner vertex of the path P_i.

Step 3. For $i = 1, \ldots, m$ each vertex of the subgraph $G(V')$ connect to the nearest inner vertex of the path P_i.

The construction of all m edge-disjoint spanning trees $T_1, \ldots, T_m$ is completed.

We denote by $W_{\mathcal{A}}$ a total weight of all trees $T_1, \ldots, T_m$ constructed by Algorithm $\mathcal{A}$. Denoting summary weights of edges, obtained on Steps 1, 2 and 3 by W_1, W_2 and W_3, we have $W_{\mathcal{A}} = W_1 + W_2 + W_3$.

Statement 1. *Running time of Algorithm $\mathcal{A}$ is $\mathcal{O}(n^2)$.*

Indeed, Preliminary Step 0 takes $\mathcal{O}(n)$ time.

On Step 1 each path is built in $\mathcal{O}(d^2)$ time, and all paths are taken $\mathcal{O}(md^2)$, or $\mathcal{O}(nd)$-running time.

On Steps 2.1–2.4 each pair (P_i, P_j), $1 \leq i < j \leq m$, of paths is connected with the edge set E_{ij} in $\mathcal{O}(d^2)$ time, and for all $\frac{m(m-1)}{2}$ pairs of paths it is required $\mathcal{O}(m^2 d^2)$, or (since $m(d + 1) \leq n$) $\mathcal{O}(n^2)$-running time.

Steps 2.5–2.6 are carried out in $\mathcal{O}(1)$ time.

Step 3 takes $\mathcal{O}(n^2)$ comparison operations.

So, the total time complexity of the Algorithm $\mathcal{A}$ is $\mathcal{O}(n^2)$ (Figs. 1, 2, and 3).

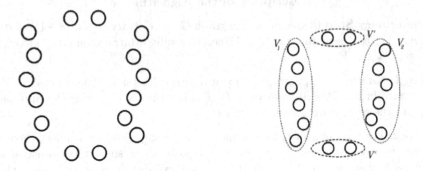

Fig. 1. Initial vertices of the graph and Step 0 of the work of the Algorithm $\mathcal{A}$ on 16-vertex complete graph, $d = 5$.

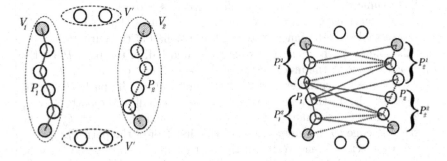

Fig. 2. Step 1 and Step 2 of the work of the Algorithm $\mathcal{A}$ on 16-vertex complete graph, $d = 5$.

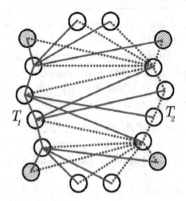

Fig. 3. Step 3 of the work of the Algorithm $\mathcal{A}$ on 16-vertex complete graph, $d = 5$.

3   A Probabilistic Analysis of Algorithm $\mathcal{A}$

We perform the probabilistic analysis under conditions that weights of graph edges are independent and identically distributed random variables η from the class of biased exponential distribution $Exp'(\lambda_n, \alpha_n)$, with density:

$$f(x) = e_{\alpha_n, \lambda_n}(x) = \begin{cases} \frac{1}{\lambda_n} e^{-\frac{x-\alpha_n}{\lambda_n}}, & \text{if } x \geq \alpha_n, \\ 0, & \text{otherwise.} \end{cases}$$

where $\lambda_n > 0, \alpha_n > 0$

By $F_A(I)$ and $OPT(I)$ we denote respectively the approximate (obtained by some approximation algorithm A) and the optimum value of the objective function of the problem on the input I. An algorithm A is said to have *estimates (performance guarantees)* $(\varepsilon_n, \delta_n)$ on the set of random inputs of the n-sized problem, if

$$\mathbf{P}\Big\{ F_A(I) > \big(1 + \varepsilon_n\big) OPT(I) \Big\} \leq \delta_n, \qquad (1)$$

where $\varepsilon_n = \varepsilon_A(n)$ is an estimation of *the relative error* of the solution obtained by algorithm A, $\delta_n = \delta_A(n)$ is an estimation of *the failure probability* of the algorithm, which is equal to the proportion of cases when the algorithm does not hold the relative error ε_n or does not produce any answer at all.

Following [5], we say that an approximation algorithm A is called *asymptotically optimal* on the class of input data of the problem, if there are exist such performance guarantees that for all input I of size n

$$\varepsilon_n \to 0 \text{ and } \delta_n \to 0 \text{ as } n \to \infty.$$

Let's introduce the following random variable $\zeta = \frac{\eta - \alpha_n}{\lambda_n}$. Note that ζ is a random variable from the class of distribution $Exp(1)$.

We denote random variable equal to minimum over k variables from the class $Exp'(\lambda_n; \alpha_n)$ (from $Exp(1)$) by η_k (ζ_k, correspondingly).

We have already mentioned that according to the description of Algorithm $\mathcal{A}$ the weight $W_{\mathcal{A}}$ of the constructed spanning tree $T_{\mathcal{A}}$ is a random value equaled of the sum $W_{\mathcal{A}} = W_1 + W_2 + W_3$. Let us consider random variables W_1, W_2, W_3:

$$W_1 = m \sum_{k=1}^{d} \eta_k = m \sum_{k=1}^{d} (\lambda_n \zeta_k + \alpha_n) \leq m(n-1)\alpha_n + \lambda_n m \sum_{k=1}^{d} \zeta_k$$

$$= m(n-1)\alpha_n + \lambda_n W_1';$$

$$W_2 = C_m^2 \left(4\tfrac{d-1}{2} \eta_{(d-1)/2} + 4\eta_{(d-1)} \right)$$

$$= C_m^2 \left(4\tfrac{d-1}{2} (\lambda_n \zeta_{(d-1)/2} + \alpha_n) + 4(\lambda_n \zeta_{(d-1)} + \alpha_n) \right)$$

$$\leq m(n-1)\alpha_n + \lambda_n C_m^2 \left(4\tfrac{d-1}{2} \zeta_{(d-1)/2} + 4\zeta_{(d-1)} \right) = m(n-1)\alpha_n + \lambda_n W_2';$$

$$W_3 = m(n - m(d+1))\eta_{(d-1)} = m(n - m(d+1))(\lambda_n \zeta_{(d-1)} + \alpha_n)$$

$$\leq m(n-1)\alpha_n + \lambda_n m(n - m(d+1))\zeta_{(d-1)} = m(n-1)\alpha_n + \lambda_n W_3'.$$

Hence we can denote the following restriction on the $W_{\mathcal{A}}$:

$$W_{\mathcal{A}} \leq m(n-1)\alpha_n + \lambda_n W'_{\mathcal{A}},$$

where

$$W'_{\mathcal{A}} = W'_1 + W'_2 + W'_3.$$

W'_1, W'_2, W'_3 are normalized random variables for values W_1, W_2, W_3, respectively.

Consider separately expectations of random variables for values W_1, W_2 and W_3.

$$\mathbf{E}W'_1 = \sum_{i=1}^{m}\sum_{k=1}^{d} \mathbf{E}\zeta_k = m\sum_{k=1}^{d}\frac{1}{k} \leq m(1+\ln d);$$

$$\mathbf{E}W'_2 = \frac{m(m-1)}{2}\left(4\frac{d-1}{2}\mathbf{E}\zeta_{(d-1)/2} + 4\mathbf{E}\zeta_{(d-1)}\right) = \frac{m(m-1)}{2}\left(4\frac{d-1}{2}\frac{2}{d-1} + \frac{4}{d-1}\right)$$

$$\leq 2\,m(m-1)\frac{d}{d-1} \leq \frac{2mn}{d-1} - 2m;$$

$$\mathbf{E}W'_3 = m(n-m(d+1))\mathbf{E}\zeta_{(d-1)} = \frac{m(n-m(d+1))}{d-1} \leq \frac{mn}{d-1} - m^2.$$

Lemma 1.

$$\mathbf{E}(W'_{\mathcal{A}}) \leq m\ln d + \frac{3mn}{d-1}.$$

Proof.

$$\mathbf{E}(W'_{\mathcal{A}}) = \mathbf{E}\left(W'_1 + W'_2 + W'_3\right) \leq m(1+\ln d) + \frac{2mn}{d-1} - 2m + \frac{mn}{d-1} - m^2 \leq m\ln d + \frac{3mn}{d-1}.$$

Lemma 2. *Algorithm $\mathcal{A}$ for solving the m-d-UMST in n-vertex complete undirected graphs with weights of edges from $Exp'(\lambda_n; \alpha_n)$ has the following estimates of the relative error ε_n and the failure probability δ_n:*

$$\varepsilon_n = (1+\omega_n)\frac{\lambda_n}{m(n-1)\alpha_n}\widehat{\mathbf{E}W'}_{\mathcal{A}}, \tag{2}$$

$$\delta_n = \mathbf{P}(\widetilde{W'}_{\mathcal{A}} > \omega_n\widehat{\mathbf{E}W'}_{\mathcal{A}}), \tag{3}$$

where $\widetilde{W'}_{\mathcal{A}} = W'_{\mathcal{A}} - \mathbf{E}W'_{\mathcal{A}}$, $\omega_n > 0$, $\widehat{\mathbf{E}W'}_{\mathcal{A}}$ is some upper bound for expectation $\mathbf{E}W'_{\mathcal{A}}$.

Proof.

$$\mathbf{P}\left\{W_{\mathcal{A}} > (1 + \varepsilon_n)OPT\right\} \leq \mathbf{P}\left\{W_{\mathcal{A}} > (1 + \varepsilon_n)m(n-1)\alpha_n\right\}$$

$$\leq \mathbf{P}\left\{m(n-1)\alpha_n + \lambda_n W'_{\mathcal{A}} > (1 + \varepsilon_n)m(n-1)\alpha_n\right\}$$

$$= \mathbf{P}\left\{W'_{\mathcal{A}} - \mathbf{E}W'_{\mathcal{A}} > \frac{\varepsilon_n m(n-1)\alpha_n}{\lambda_n} - \mathbf{E}W'_{\mathcal{A}}\right\}$$

$$= \mathbf{P}\left\{\widetilde{W}'_{\mathcal{A}} > \frac{\varepsilon_n m(n-1)\alpha_n}{\lambda_n} - \mathbf{E}W'_{\mathcal{A}}\right\}$$

$$\leq \mathbf{P}\left\{\widetilde{W}'_{\mathcal{A}} > \frac{\varepsilon_n m(n-1)\alpha_n}{\lambda_n} - \widehat{\mathbf{E}W}'_{\mathcal{A}}\right\} = \mathbf{P}\left\{\widetilde{W}'_{\mathcal{A}} > \omega_n \widehat{\mathbf{E}W}'_{\mathcal{A}}\right\} = \delta_n.$$

Further for the probabilistic analysis of Algorithm $\mathcal{A}$ we use the following probabilistic statement

Petrov's Theorem [8]. *Consider independent random variables $X_1, \ldots, X_n$. Let there be positive constants T and $h_1, \ldots, h_n$ such that for all $k = 1, \ldots, n$ and $0 \leq t \leq T$ the following inequalities hold:*

$$\mathbf{E}e^{tX_k} \leq \exp\left\{\frac{h_k t^2}{2}\right\}. \tag{4}$$

Set $S = \sum_{k=1}^{n} X_k$ and $H = \sum_{k=1}^{n} h_k$. Then

$$\mathbf{P}\{S > x\} \leq \begin{cases} \exp\left\{-\frac{x^2}{2H}\right\}, & \text{if } 0 \leq x \leq HT, \\ \exp\left\{-\frac{Tx}{2}\right\}, & \text{if } x \geq HT. \end{cases}$$

Lemma 3. *Let $t \geq 0$. Then:*

$$\mathbf{E}e^{t\zeta_k} = \frac{1}{1 - t/k}. \tag{5}$$

Proof. Note that the random variable ζ_k is from the class $Exp(k)$. Then consider a distribution function $G(x)$ of $e^{t\zeta_k}$:

$$G(x) = \mathbf{P}(e^{t\zeta_k} \leq x) = \mathbf{P}(t\zeta_k \leq \ln x) = \mathbf{P}\left(\zeta_k \leq \frac{\ln x}{t}\right) = \begin{cases} 1 - \exp\left\{-k\frac{\ln x}{t}\right\}, & \text{if } x \geq 1, \\ 0, & \text{if } x < 1. \end{cases}$$

$$= \begin{cases} 1 - x^{-\frac{k}{t}}, & \text{if } x \geq 1, \\ 0, & \text{if } x < 1. \end{cases}$$

Distribution function above is correct for every $k \geq 1$.

Now let's calculate density $g(x)$:

$$g(x) = G'(x) = \begin{cases} \frac{k}{t} x^{-(\frac{k}{t}+1)}, & \text{if } x \geq 1, \\ 0, & \text{if } x < 1. \end{cases}$$

Denote $q = \frac{k}{t}$. Then:

$$\mathbf{E}e^{t\zeta_k} = \int_1^{+\infty} xqx^{-(q+1)}dx = \int_1^{+\infty} qx^{-q}dx = \frac{q}{q-1} = \frac{1}{1-t/k}.$$

Lemma 4. *Let ζ_k be a random variable, which is equal to minimum over k variables from the class $Exp(1)$. We also define $T = 3/4$ and $h_k = \frac{7}{k^2}$. Therefore random variables $\tilde{\zeta}_k = \zeta_k - \mathbf{E}\zeta_k$ satisfy requirement (4) of Petrov's theorem for every $0 \leq t \leq T$ and $1 \leq k \leq n$.*

Proof. Note that $\mathbf{E}\zeta_k = \frac{1}{k}$. Let's consider an above bound of expectation $\mathbf{E}e^{t\zeta_k}$ taking into account that $\frac{t}{k} \leq \frac{3}{4}$:

$$\mathbf{E}e^{t\zeta_k} = \frac{1}{1-t/k} = \sum_{m=0}^{\infty} \left(\frac{t}{k}\right)^m \leq 1 + \frac{t}{k} + \left(\frac{t}{k}\right)^2 \frac{1}{1-t/k} \leq 1 + \frac{t}{k} + 4\left(\frac{t}{k}\right)^2$$

$$= 1 + \frac{t}{k} + \frac{1}{2}\left(\frac{t}{k}\right)^2 + \frac{7}{2}\left(\frac{t}{k}\right)^2 \leq \left(1 + \frac{t}{k} + \frac{1}{2}\left(\frac{t}{k}\right)^2\right)\left(1 + \frac{7}{2}\left(\frac{t}{k}\right)^2\right)$$

$$\leq e^{\frac{t}{k}}e^{\frac{7t^2}{2k^2}} = e^{t\mathbf{E}\zeta_k}e^{\frac{h_k t^2}{2}}.$$

Lemma 5. *Let $\ln n \leq d < \frac{n}{m}$, $m \geq 2$ and n such as $n > 10$. Then following upper bound is correct for sum of constants $h_k = \frac{7}{k^2}$, corresponding to edges of trees T_s, $s \in 1, ..., m$*

$$H \leq \frac{7mn}{d-1}.$$

Proof. Let H be the sum of H_1, H_2 and H_3 according to the steps of Algorithm $\mathcal{A}$ number 1, 2, 3 respectively. From this and from estimate of above we get:

$$H_1 = m\sum_{k=1}^{d} h_k = m\sum_{k=1}^{d} \frac{7}{k^2} \leq 7\psi m,$$

where $\psi = 1.645$. Here we use a well known estimate reached by Euler $1 + \frac{1}{2^2} + \frac{1}{3^2} + \frac{1}{4^2} + ... = \frac{\pi^2}{6} < 1.645$.

$$H_2 = C_m^2\left(4\frac{d-1}{2}h_{(d-1)/2} + 4h_{(d-1)}\right) = 7\frac{m(m-1)}{2}4\left(\frac{d-1}{2}\frac{4}{(d-1)^2} + \frac{1}{(d-1)^2}\right)$$

$$\leq 7 \cdot 2m^2\left(\frac{2}{d-1} + \frac{1}{(d-1)^2}\right) \leq 7\frac{4m^2 d}{(d-1)^2};$$

$$H_3 = m(n - m(d+1))h_{(d-1)} = m(n - m(d+1))\frac{7}{(d-1)^2} \leq 7\frac{mn}{(d-1)^2} - 7m^2\frac{d+1}{(d-1)^2}.$$

Therefore:

$$H = H_1 + H_2 + H_3 = 7\psi m + 7\frac{4m^2 d}{(d-1)^2} + 7\frac{mn}{(d-1)^2} - 7m^2\frac{d+1}{(d-1)^2}$$

$$\leq 7\left(\psi m + \frac{3m^2 d}{(d-1)^2} + \frac{mn}{(d-1)^2}\right) \leq 7\left(\psi m + \frac{4mn}{(d-1)^2}\right) = \frac{7mn}{d-1}\left(\frac{\psi(d-1)}{n} + \frac{4}{d-1}\right).$$

Notice following bound for the diameter $d \leq \frac{n}{m} \leq \frac{n}{2}$. Then we can estimate the parameter H as follows:

$$H \leq \frac{7mn}{d-1}\left(\frac{\psi n/2}{n} + \frac{4}{n/2-1}\right) \leq \frac{7mn}{d-1}\left(\frac{\psi}{2} + \frac{8}{n-2}\right) \leq \frac{7mn}{d-1}.$$

Lemma 6.
$$\widehat{EW'}_{\mathcal{A}} = \frac{7mn}{d-1}.$$

Proof. Knowing that $\ln d \leq \ln n < \frac{n}{d}$, we have:

$$EW'_{\mathcal{A}} \leq m\ln d + \frac{3mn}{d-1} < m\frac{n}{d} + \frac{3mn}{d-1} < \frac{4mn}{d-1} < \frac{7mn}{d-1}.$$

Lemma 6 is proved.

Lemma 7.
$$\widehat{EW'}_{\mathcal{A}} = 12m\ln n.$$

Proof. Knowing that $\ln d \leq \ln n$, $n \leq d\ln n$, we have:

$$EW'_{\mathcal{A}} \leq m\ln d + \frac{3mn}{d-1} \leq m\ln n + 3m\ln n\frac{d}{d-1}.$$

Since $d \geq 2$ we have $\frac{d}{d-1} \leq 2$.
Therefore:
$$EW'_{\mathcal{A}} \leq 7m\ln n < 12m\ln n.$$

Lemma 7 is proved.

Theorem 1. *Let the parameter $d = d_n$ be defined so that*

$$\ln n \leq d < \frac{n}{m}, \tag{6}$$

Then Algorithm $\mathcal{A}$ solves the problem m-d-UMST in n-vertex complete graphs with weights of edges from $Exp'(\lambda_n; \alpha_n)$ with the failure probability

$$\delta_n = n^{-m} \to 0, \ as \ n \to \infty, \tag{7}$$

and the following conditions of asymptotical optimaity

$$\frac{\lambda_n}{\alpha_n} = \begin{cases} o(d), & if \ \ln n \leq d < \frac{n}{\ln n}, \\ o(\frac{n}{\ln n}), & if \ \frac{n}{\ln n} \leq d < \frac{n}{m} \ and \ m < \ln n. \end{cases} \tag{8}$$

Proof. Firstly, let us note that during the work of the Algorithm $\mathcal{A}$ we have a deal with random variables like $\zeta_k, 1 \leq k \leq d$. This variables satisfy the conditions of Petrov's Theorem with constants $T = \frac{3}{4}$ and $h_k = \frac{7}{k^2}$.

We will carry out a proof for two cases of possible values of the parameter d: $\ln n \leq d < \frac{n}{\ln n}$ and $\frac{n}{\ln n} \leq d < \frac{n}{m}$.

Case 1: $\ln n \leq d < \frac{n}{\ln n}$

Using the formula (2) and Lemma 6 we have:

$$\varepsilon_n = (1 + w_n)\frac{\lambda_n}{m(n-1)\alpha_n}\widehat{EW'}_{\mathcal{A}} = (1 + w_n)\frac{\lambda_n}{m(n-1)\alpha_n}\frac{4mn}{d-1}$$

$$= (1 + w_n)\frac{4n}{n-1}\frac{(\lambda_n/\alpha_n)}{d-1}.$$

Setting $w_n = 1$ we see that $\varepsilon_n \to 0$ under the condition:

$$\frac{\lambda_n}{\alpha_n} = o(d).$$

Now using Petrov's Theorem and Lemma 6 we consider a failure probability:

$$\delta_n = \mathbf{P}(\widetilde{W'}_{\mathcal{A}} > w_n\widehat{EW'}_{\mathcal{A}}) = \mathbf{P}(\widetilde{W'}_{\mathcal{A}} > \frac{7mn}{d-1}).$$

Define constants $T = \frac{3}{4}$ and $h_k = \frac{7}{k^2}$ for edges, whose weight corresponds to a random variable ζ_k, and which are included in the spanning trees $T_{\mathcal{A}}$.

From the Lemma 5 we have:

$$TH = \frac{3}{4} \cdot \frac{7mn}{d-1} < \frac{7mn}{d-1} = x.$$

In accordance with the Petrov's Theorem we get following estimation:

$$\delta_n = \mathbf{P}(\widetilde{W'}_{\mathcal{A}} > x) \leq \exp\left\{-\frac{Tx}{2}\right\}.$$

Since $n > d \ln n$ we have:

$$\frac{Tx}{2} = \frac{3}{8} \cdot \frac{7mn}{d-1} > \frac{21}{8}\frac{md\ln n}{d-1} > m\ln n.$$

Hence:

$$\delta_n \leq \exp\left\{-\frac{Tx}{2}\right\} \leq n^{-m} \to 0.$$

Therefore in case $\ln n \leq d < \frac{n}{\ln n}$ the Algorithm $\mathcal{A}$ solves the problem m-d-UMST in n-vertex complete undirected graph with weights of edges from $Exp'(\lambda_n, \alpha_n)$ asymptotically optimal.

Case 2: $\frac{n}{\ln n} \leq d < \frac{n}{m}$

Using the formula (2) and Lemma 7 we have:

$$\varepsilon_n = (1 + w_n)\frac{\lambda_n}{m(n-1)\alpha_n}\widehat{EW'}_{\mathcal{A}} = (1 + w_n)\frac{\lambda_n}{m(n-1)\alpha_n}12m\ln n$$

$$= 12(1 + w_n)\frac{\lambda_n/\alpha_n}{(n-1)/\ln n}.$$

Setting $\omega_n = 1$ we see that $\varepsilon_n \to 0$ under the condition:

$$\frac{\lambda_n}{\alpha_n} = o\left(\frac{n}{\ln n}\right).$$

Now using Petrov's Theorem and Lemma 7 we consider a failure probability:

$$\delta_n = \mathbf{P}(\widetilde{W}'_{\mathcal{A}} > \omega_n \widehat{\mathbf{E}W'}_{\mathcal{A}}) = \mathbf{P}(\widetilde{W}'_{\mathcal{A}} > 12m \ln n).$$

Define constants $T = \frac{3}{4}$ and $h_k = \frac{7}{k^2}$ for edges, whose weight corresponds to a random variable ζ_k, and which are included in the spanning trees $T_{\mathcal{A}}$. From the Lemma 5 and the condition $n \le d \ln n$ we have:

$$TH = \frac{3}{4} \cdot \frac{7mn}{d-1} < \frac{3}{4} \cdot 7m \ln n \frac{d}{d-1} \le \frac{3}{2} \cdot 7m \ln n < 12\,m \ln n = x.$$

In accordance with the Petrov's Theorem we get following estimation:

$$\delta_n = \mathbf{P}(\widetilde{W}'_{\mathcal{A}} > x) \le \exp\left\{-\frac{Tx}{2}\right\}.$$

Then:

$$\frac{Tx}{2} = \frac{3}{8} \cdot 12m \ln n > m \ln n.$$

Finally:

$$\delta_n \le \exp\left\{-\frac{Tx}{2}\right\} \le n^{-m} \to 0.$$

Therefore in the second case $\frac{n}{\ln n} \le d < \frac{n}{m}$ the Algorithm $\mathcal{A}$ also solves the problem m-d-UMST in n-vertex complete graph with weights of edges from $Exp'(\lambda_n, \alpha_n)$ asymptotically optimal.

We conclude that within the values of parameter d for both cases we have estimated of the relative error $\varepsilon_n \to 0$ and the failure probability $\delta_n \to 0$ as $n \to \infty$.
Theorem 1 is completely proved.

4   Conclusion

In current work, we have carried out the probabilistic analysis of our algorithm from [7] for the m-d-UMST problem in the case of biased exponential distribution of edge weights of complete undirected graph. We have proved that this algorithm solves the m-d-UMST asymptotically optimal for such distribution and the conditions of asymptotic optimality for described algorithm have been presented. For further investigations in this area it would be interesting to perform probabilistic analysis for the m-d-UMST in the case of truncated-normal distribution as well as discrete distributions. Also it will be intriguing to consider related problem of finding Maximum Spanning Trees in complete undirected graph.

References

1. Angel, O., Flaxman, A.D., Wilson, D.B.: A sharp threshold for minimum bounded-depth and bounded-diameter spanning trees and Steiner trees in random networks. Combinatorica **32**, 1–33 (2012). https://doi.org/10.1007/s00493-012-2552-z
2. Cooper, C., Frieze, A., Ince, N., Janson, S., Spencer, J.: On the length of a random minimum spanning tree. Combinator. Probab. Comput. **25**(1), 89–107 (2016)
3. Frieze, A.: On the value of a random MST problem. Discrete Appl. Math. **10**, 47–56 (1985)
4. Garey, M.R., Johnson, D.S.: Computers and Intractability, 340 p. Freeman, San Francisco (1979)
5. Gimadi, E.Kh., Glebov, N.I., Perepelitsa, V.A.: Algorithms with estimates for discrete optimization problems. Problemy Kibernetiki **31**, 35–42 (1975). (in Russian)
6. Gimadi, E.K., Shevyakov, A.S., Shtepa, A.A.: A given diameter MST on a random graph. In: Olenev, N., Evtushenko, Y., Khachay, M., Malkova, V. (eds.) OPTIMA 2020. LNCS, vol. 12422, pp. 110–121. Springer, Cham (2020). https://doi.org/10.1007/978-3-030-62867-3_9
7. Gimadi, E.K., Shevyakov, A.S., Shtepa, A.A.: On Asymptotically Optimal Approach for the Problem of Finding Several Edge-Disjoint Spanning Trees of Given Diameter in an Undirected Graph with Random Edge Weights, In: Pardalos P., Khachay M., Kazakov A. (eds) Mathematical Optimization Theory and Operations Research, MOTOR 2021, LNCS, 12755, Springer, Cham (2021). https://doi.org/10.1007/978-3-030-77876-7_5
8. Petrov, V.V.: Limit theorems of probability theory. In: Sequences of Independent Random Variables, 304 p. Clarendon Press, Oxford (1995)

Neural Architecture Search
with Structure Complexity Control

Konstantin D. Yakovlev[1]([✉]), Olga S. Grebenkova[1], Oleg Y. Bakhteev[1,2],
and Vadim V. Strijov[1,2]

[1] MIPT, Dolgoprudny, Russia
{iakovlev.kd,grebenkova.os,bakhteev,strijov}@phystech.edu
[2] Dorodnicyn Computing Center RAS, Moscow, Russia

Abstract. The paper investigates the problem of deep learning model selection. We propose a method of a neural architecture search with respect to the desired model complexity called DARTS-CC. An amount of parameters in the model is considered as a model complexity. The proposed method is based on a differential architecture search algorithm (DARTS). Instead of optimizing structural parameters of the architecture, we consider them as a function depending on the complexity parameter. It enables us to obtain multiple architectures at one optimization procedure and select the architecture based on our computation budget. To evaluate the performance of the proposed algorithm, we conduct experiments on the Fashion-MNIST and CIFAR-10 datasets and compare the resulting architecture with architectures obtained by other neural architecture search methods.

Keywords: Differential architecture search · Deep learning · Hypernetwork · Neural networks · Model complexity control

1 Introduction

In this paper, we consider the problem of searching the architecture of a deep learning model with the control of its complexity. A model is considered as a directed graph with edges corresponding to non-linear functions on the dataset and the vertices corresponding to intermediate representations of the dataset under the operations. In this paper, we base on the differential algorithm

This paper contains results of the project Mathematical methods of intelligent big data analysis, which is carried out within the framework of the Program "Center of Big Data Storage and Analysis" of the National Technology Initiative Competence Center. It is supported by the Ministry of Science and Higher Education of the Russian Federation according to the agreement between the M.V. Lomonosov Moscow State University and the Foundation of project support of the National Technology Initiative from 11.12.2018, No 13/1251/2018. This research was supported by RFBR (projects 19-07-01155, 19-07-00885).

E. Burnaev et al. (Eds.): AIST 2021, CCIS 1573, pp. 207–219, 2022.
https://doi.org/10.1007/978-3-031-15168-2_17

DARTS [16]. It solves the problem of searching the model architecture by translating the search space of structural parameters from a discrete to a continuous representation. In contrast to previous algorithms that were based on discrete optimisation problem, DARTS does not suffer from combinatorial explosion. Due to relaxation, the algorithm does not use an exhaustive search to solve the optimization task. Furthermore, this relaxation enables us to use gradient-based optimization for the model architecture selection.

Although a significant success in neural architecture search-based model selection [25], the problem of model selection is still can be a challenge, especially when dealing with constraints on computational budget or model size [11,21]. The paper [21] presents an approach to use regularization term to control the model complexity. It mainly focuses on the target hardware for further model exploitation. A similar approach can be found in the paper [11], where a neural architecture search is employed with a limited resource (RC-DARTS). Restrictions are added to the basic DARTS algorithm, such as the number of model parameters. In order to solve the problem of conditional optimization, an iterative projection algorithm is introduced, which consists of the fact that after a certain number of iterations of gradient descent, the projection occurs on the set specified by the constraints. However, both algorithms require the launch of an individual architecture search process for every set of complexity values, so they still can be too time-consuming.

This paper investigates the problem of the model complexity control during architecture search. Opposite to the listed works above, our method, which is called DARTS-CC, is based on the hypernetwork concept [8]. A hypernetwork is a model that generates the parameters of the target deep learning model. In this paper, we employ hypernetworks to generate structural parameters that control the final architecture of the model. It enables us to obtain multiple neural architectures during the architecture search procedure for further model fine-tuning with an architecture that suits our computation budget in the best way. The main idea of our paper and the difference between our method and DARTS is shown in Fig. 1. Instead of using constant structural parameters $\alpha^{(i,j)}$ that control the model architecture, we propose to consider them as functions on the model complexity parameter λ. We also use Gumbel-softmax distribution for sampling non-linear operations instead of using softmax function on the structural parameters $\alpha^{(i,j)}$: this gives us the relaxed architecture very close to the target discrete one. To the best of our knowledge, this is the first paper that considers searching the family of models at once and then choosing the suitable complexity value and corresponding model at the inference step. So we do not need to restart the architecture search for models with different complexities. The computational experiment is performed on the Fashion-MNIST [22] and CIFAR-10 [12] dataset.

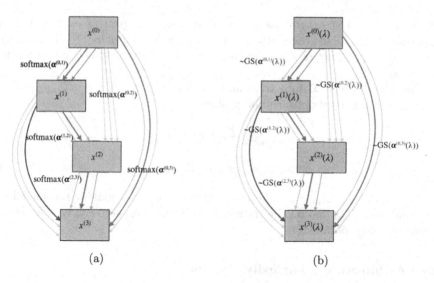

Fig. 1. An overview of DARTS (a) and proposed method DARTS-CC (b).

2   Problem Statement

Below we briefly describe the DARTS method and our approach based on it. Given a classification problem with a dataset $\mathfrak{D} = (\mathbf{X}, \mathbf{y})$. For each object $\mathbf{x} \in \mathbf{X}$ there is a label $y \in \mathbf{y}$.

The approach presented in DARTS considers a neural architecture as a sequence of repeated submodels, called *cells*. Each cell has an identical structure but has different parameter values [25]. It is represented as a directed graph. The graph consists of vertices and edges between them. Formally, there is a set of vertices $V = \{1, \ldots, N\}$ and a set of edges $E = \{(i, j) \in V \times V \mid i < j\}$, where N is the number of vertices. For each edge (i, j) between vertices i and j there is a non-linear function, called an *operation* $\mathbf{g}^{(i,j)}$ from the vector $\mathbf{g}^{(i,j)}$ of all the available non-linear functions for the target edge (i, j). The values in each of the intermediate nodes $\mathbf{x}^{(j)}$ are defined through the values in the nodes with the lower number i:

$$\mathbf{x}^{(j)} = \sum_{(i,j) \in E} \mathbf{g}^{(i,j)}(\mathbf{x}^{(i)}), \qquad (1)$$

where $\mathbf{x} = \mathbf{x}^{(0)}$.

The task of architecture search is to choose the non-linear operations $\mathbf{g}^{(i,j)}$ for each edge in the cell. In order to reduce the discrete optimization problem to the continuous optimization problem introduce a mixed operation for each edge (i, j):

$$\hat{\mathbf{g}}^{(i,j)}(\mathbf{x}^{(i)}) = \langle \mathbf{softmax}(\boldsymbol{\alpha}^{(i,j)}), \mathbf{g}^{(i,j)}(\mathbf{x}^{(i)}) \rangle, \qquad (2)$$

where $\boldsymbol{\alpha}^{(i,j)}$ is a structural parameter that assigns impact of each non-linear operation $\mathbf{g}^{(i,j)}$. Thus, each edge (i,j) is assigned a vector $\boldsymbol{\alpha}^{(i,j)}$ of dimension equal to $\dim \mathbf{g}^{(i,j)}$. Let $\boldsymbol{\alpha} \in \mathbb{R}^s$ be a concatenation of all the structural parameter vectors $\boldsymbol{\alpha}^{(i,j)}$.

Split the dataset $\mathfrak{D}$ into train and validation parts: $\mathfrak{D} = \mathfrak{D}_{\text{train}} \sqcup \mathfrak{D}_{\text{val}}$. Formulate a two-level optimization problem:

$$\min_{\boldsymbol{\alpha} \in \mathbb{R}^s} \mathcal{L}_{\text{val}}(\mathbf{w}^*, \boldsymbol{\alpha}), \tag{3}$$

$$\text{s.t.} \quad \mathbf{w}^* = \arg \min_{\mathbf{w} \in \mathbb{R}^n} \mathcal{L}_{\text{train}}(\mathbf{w}, \boldsymbol{\alpha}). \tag{4}$$

Here $\mathcal{L}_{\text{val}}$ and $\mathcal{L}_{\text{train}}$ are the cross-entropy loss functions of the model on the validation $\mathfrak{D}_{\text{val}}$ and on the train dataset $\mathfrak{D}_{\text{train}}$, respectively, $\mathbf{w} \in \mathbb{R}^n$ is the vector of model parameters.

2.1   Architecture Complexity Control

Our modification of the DARTS method consists of two parts. First, similar to [3] instead of using softmax operation in mixing operations $\hat{\mathbf{g}}^{(i,j)}$ we employ Gumbel-softmax distribution. We sample edge values at each training step from it:

$$\hat{\mathbf{g}}^{(i,j)}(\mathbf{x}^{(i)}) = \langle \boldsymbol{\gamma}^{(i,j)}, \mathbf{g}^{(i,j)}(\mathbf{x}^{(i)}) \rangle, \quad \boldsymbol{\gamma}^{(i,j)} \sim \mathcal{GS}(\exp(\boldsymbol{\alpha}^{(i,j)}), t), \tag{5}$$

where $\mathcal{GS}$ is a Gumbel-softmax distribution [10], t controls the temperature of the distribution. With $t \to 0$ the distribution tends to be similar to the discrete, which allow us to obtain a relaxed version of the architecture close to the discrete one.

Define an *architecture complexity* as an amount of parameters that the corresponding model has. In order to control it modify the loss function $\mathcal{L}_{\text{val}}$:

$$\mathcal{L}'_{\text{val}} = \mathsf{E}_{\boldsymbol{\gamma}} \left(\mathcal{L}_{\text{val}} + \lambda \sum_{(i,j) \in E} \langle \boldsymbol{\gamma}^{(i,j)}, \mathbf{n}(\mathbf{g}^{(i,j)}) \rangle \right), \tag{6}$$

where $\boldsymbol{\gamma}$ is a concatenation of all the $\boldsymbol{\gamma}^{(i,j)}$ vectors, $(i,j) \in E$, $\mathbf{n}$ is a vector of numbers of parameters for each operation $\mathbf{g}^{(i,j)} \in \mathbf{g}^{(i,j)}$, λ is a regularization coefficient.

The following theorem shows that we get an optimization close to the discrete one when the temperature tends to zero.

Theorem 1. *Let for each* $\mathbf{x} \in \mathbf{X}$, *for each* $\mathbf{w} \in \mathbb{R}^n$ *the function* $\mathcal{L}_{\mathrm{val}}(\mathbf{w}, \boldsymbol{\gamma})$ *is continuous by* $\boldsymbol{\gamma}$. *Then the following expression is true:*

$$
\lim_{t \to +0} \mathsf{E}_{\boldsymbol{\gamma}} \left(\mathcal{L}_{\mathrm{val}}(\mathbf{w}, \boldsymbol{\gamma}) + \lambda \sum_{(i,j) \in E} \langle \boldsymbol{\gamma}^{(i,j)}, \mathbf{n}(\mathbf{g}^{(i,j)}) \rangle \right)
$$

$$
= \mathsf{E}_{\tilde{\boldsymbol{\gamma}}} \left(\mathcal{L}_{\mathrm{val}}(\mathbf{w}, \tilde{\boldsymbol{\gamma}}) + \lambda \sum_{(i,j) \in E} \langle \tilde{\boldsymbol{\gamma}}^{(i,j)}, \mathbf{n}(\mathbf{g}^{(i,j)}) \rangle \right)
$$

$$
= \sum_{1 \le k^{(i,j)} \le \dim \mathbf{g}^{(i,j)}, \ (i,j) \in E} \mathcal{L}_{\mathrm{val}}(\mathbf{w}, [\mathbf{e}^{(i,j)}(k^{(i,j)})]) \prod_{(l,m) \in E} \mathrm{softmax}(\boldsymbol{\alpha}^{(l,m)})_{k^{(i,j)}}
$$

$$
+ \lambda \sum_{(i,j) \in E} \langle \mathrm{softmax}(\boldsymbol{\alpha}^{(i,j)}), \mathbf{n}(\mathbf{g}^{(i,j)}) \rangle, \quad (7)
$$

where $\mathbf{e}^{(i,j)}$ *is a one-hot vector:*

$$
\mathbf{e}^{(i,j)}(k) \in \mathbb{R}^{\dim \mathbf{g}^{(i,j)}}, \quad 1 \le k \le \dim \mathbf{g}^{(i,j)}, \quad e_k^{(i,j)}(k) = 1, \quad e_m^{(i,j)}(k) = 0, m \neq k,
$$

vector $[\mathbf{e}^{(i,j)}(k^{(i,j)})]$ *is a concatenation of vectors* $\mathbf{e}^{(i,j)}(k^{(i,j)})$ *over all edges* $(i,j) \in E$, *vector* $\tilde{\boldsymbol{\gamma}}$ *is a concatenation of the following multinomially distributed variables* $\tilde{\boldsymbol{\gamma}}^{(i,j)}$:

$$
p\left(\tilde{\boldsymbol{\gamma}}^{(i,j)} = \mathbf{e}^{(i,j)}(k) \right) = \frac{\exp(\alpha_k^{(i,j)})}{\sum_{m=1}^{\dim \mathbf{g}^{(i,j)}} \exp(\alpha_m^{(i,j)})} = \mathrm{softmax}(\boldsymbol{\alpha}^{(i,j)})_k. \quad (8)
$$

In our notation $\mathrm{softmax}(\boldsymbol{\alpha}^{(i,j)})_k$ *is the* k-*th component of the vector* $\mathrm{softmax}(\boldsymbol{\alpha}^{(i,j)})$.

Proof. From zero temperature property of Gumbel-Softmax distribution [18] it follows that:

$$
\boldsymbol{\gamma}^{(i,j)} \xrightarrow[t \to +0]{\mathrm{a.s.}} \tilde{\boldsymbol{\gamma}}^{(i,j)},
$$

Then, by the Mann-Wald theorem:

$$
\mathcal{L}_{\mathrm{val}}(\mathbf{w}, \boldsymbol{\gamma}) \xrightarrow[t \to +0]{\mathrm{a.s.}} \mathcal{L}_{\mathrm{val}}(\mathbf{w}, \tilde{\boldsymbol{\gamma}}),
$$

$$
\langle \boldsymbol{\gamma}^{(i,j)}, \mathbf{n}(\mathbf{g}^{(i,j)}) \rangle \xrightarrow[t \to +0]{\mathrm{a.s.}} \langle \tilde{\boldsymbol{\gamma}}^{(i,j)}, \mathbf{n}(\mathbf{g}^{(i,j)}) \rangle.
$$

Since the vector $\boldsymbol{\gamma}^{(i,j)}$ is an element of a simplex, then $\boldsymbol{\gamma}$ is an element of a compact set. According to the extreme value theorem, $|\mathcal{L}_{\mathrm{val}}(\mathbf{w}, \boldsymbol{\gamma})|$ is bounded on it. Similarly, $|\langle \boldsymbol{\gamma}^{(i,j)}, \mathbf{n}(\mathbf{g}^{(i,j)}) \rangle|$ as a function of $\boldsymbol{\gamma}^{(i,j)}$ is also bounded on it. Using dominated convergence theorem, we can swap mathematical expectation and limit:

$$
\lim_{t \to +0} \mathsf{E}_{\boldsymbol{\gamma}} \mathcal{L}_{\mathrm{val}}(\mathbf{w}, \boldsymbol{\gamma}) = \mathsf{E}_{\tilde{\boldsymbol{\gamma}}} \mathcal{L}_{\mathrm{val}}(\mathbf{w}, \tilde{\boldsymbol{\gamma}}), \quad (9)
$$

$$
\lim_{t \to +0} \mathsf{E}_{\boldsymbol{\gamma}^{(i,j)}} \langle \boldsymbol{\gamma}^{(i,j)}, \mathbf{n}(\mathbf{g}^{(i,j)}) \rangle = \mathsf{E}_{\tilde{\boldsymbol{\gamma}}^{(i,j)}} \langle \tilde{\boldsymbol{\gamma}}^{(i,j)}, \mathbf{n}(\mathbf{g}^{(i,j)}) \rangle. \quad (10)
$$

Write down the mathematical expectation (9) in explicit form, using (8):

$$
\begin{aligned}
\mathsf{E}_{\tilde{\gamma}} \, & \mathcal{L}_{\mathrm{val}}(\mathbf{w}, \tilde{\gamma}) \\
&= \sum_{1 \leq k^{(i,j)} \leq \dim \mathbf{g}^{(i,j)}, \ (i,j) \in E} \mathcal{L}_{\mathrm{val}}(\mathbf{w}, [\mathbf{e}^{(i,j)}(k^{(i,j)})]) p(\tilde{\gamma} = [\mathbf{e}^{(i,j)}(k^{(i,j)})]) \\
&= \sum_{1 \leq k^{(i,j)} \leq \dim \mathbf{g}^{(i,j)}, \ (i,j) \in E} \mathcal{L}_{\mathrm{val}}(\mathbf{w}, [\mathbf{e}^{(i,j)}(k^{(i,j)})]) \prod_{(l,m) \in E} \mathrm{softmax}(\boldsymbol{\alpha}^{(l,m)})_{k^{(i,j)}}. \quad (11)
\end{aligned}
$$

Do the same with the mathematical expectation (10):

$$
\begin{aligned}
\mathsf{E}_{\tilde{\gamma}^{(i,j)}} \langle \tilde{\gamma}^{(i,j)}, \mathbf{n}(\mathbf{g}^{(i,j)}) \rangle &= \langle \mathsf{E}_{\tilde{\gamma}^{(i,j)}} \tilde{\gamma}^{(i,j)}, \mathbf{n}(\mathbf{g}^{(i,j)}) \rangle \\
&= \langle \mathrm{softmax}(\boldsymbol{\alpha}^{(i,j)}), \mathbf{n}(\mathbf{g}^{(i,j)}) \rangle. \quad (12)
\end{aligned}
$$

From (9), (10), (11), and (12) follows (7).

To reduce stochasticity in our experiments we use a loss function with simplified regularization term:

$$
\mathcal{L}'_{\mathrm{val}} = \mathcal{L}_{\mathrm{val}} + \lambda \sum_{(i,j) \in E} \langle \mathrm{softmax}(\gamma^{(i,j)}), \mathbf{n}(\mathbf{g}^{(i,j)}) \rangle, \quad (13)
$$

Note that both the original loss function (6) and its modified version (13) become equivalent to the discrete optimization with $t \to +0$.

In order to control the model complexity after the structure optimization, we employ the concept of hypernetwork. Let Λ be a set of the regularization parameter λ values. A hypernetwork is a parametric mapping from the set Λ to the set of model's structural parameters $\mathbb{R}^s$ [7]:

$$
\boldsymbol{\alpha} : \Lambda \times \mathbb{R}^u \to \mathbb{R}^s,
$$

where $\mathbb{R}^u$ hypernetwork parameter space.

Instead of using a fixed structural parameters vector $\boldsymbol{\alpha}^{(i,j)}$, we redefine it as a hypernetwork:

$$
\boldsymbol{\alpha}^{(i,j)} = \boldsymbol{\alpha}^{(i,j)}(\lambda, \mathbf{a}^{(i,j)}), \quad \lambda \in \Lambda,
$$

where $\mathbf{a}^{(i,j)}$ is a vector of the parameters of the $\boldsymbol{\alpha}^{(i,j)}$ function.

In this paper each function $\boldsymbol{\alpha}^{(i,j)}, (i,j) \in E$ is a piecewise linear function:

$$
\begin{aligned}
\boldsymbol{\alpha}^{(i,j)}(\lambda, \mathbf{a}^{(i,j)}) = \sum_{k=0}^{N-1} \Bigg(& \frac{\lambda - r_k}{r_{k+1} - r_k} a_k^{(i,j)} \\
&+ \left(1 - \frac{\lambda - r_k}{r_{k+1} - r_k}\right) a_{k+1}^{(i,j)} \Bigg) I \left[\lambda \in [r_k, r_{k+1}]\right], \quad (14)
\end{aligned}
$$

where $a_k^{(i,j)}$ are the parameters of the function, $r_k \in \mathbb{R}$ are the limits of each linear part of the function, I is an indicator function. For better model fitting, we also use a hypernetwork piecewise linear layer for the last model layer.

The final optimization function is the following:

$$\min_{a \in \mathbb{R}^u} \mathsf{E}_{\lambda \in p(\lambda)} \bigg(\mathsf{E}_{\gamma^{(i,j)} \sim \mathcal{GS}(\alpha^{(i,j)}, \lambda)} \mathcal{L}_{\text{val}}\left(\mathbf{w}^*, \gamma\right)$$

$$+ \lambda \sum_{(i,j) \in E} \langle \mathbf{softmax}(\alpha^{(i,j)}(\lambda, \mathbf{a})), \mathbf{n}(\mathbf{g}^{(i,j)}) \rangle \bigg), \quad (15)$$

$$\text{s.t.} \quad \mathbf{w}^* = \arg \min_{\mathbf{w} \in \mathbb{R}^n} \mathsf{E}_{\lambda \in p(\lambda)} \bigg(\mathsf{E}_{\gamma^{(i,j)} \sim \mathcal{GS}(\alpha^{(i,j)}, \lambda)} \mathcal{L}_{\text{train}}\left(\mathbf{w}, \gamma\right) \bigg), \quad (16)$$

where $p(\lambda)$ is a predefined distribution over Λ, $\mathbf{a} \in \mathbb{R}^u$ is a concatenation of all the parameters of the functions $\alpha^{(i,j)}$, γ is a concatenation of all vectors $\gamma^{(i,j)}$. The optimization algorithm is shown in Fig. 2.

Algorithm 1 DARTS-CC

1: Initialize $\mathbf{a} \in \mathbb{R}^u$, $\mathbf{w} \in \mathbb{R}^n$
2: **while** not converged **do**
3:     Sample $\lambda \sim p(\lambda)$
4:     Sample $\gamma^{(i,j)} \sim \mathcal{GS}(\alpha^{(i,j)}, \lambda)$ for each $(i,j) \in E$
5:     Update $\mathbf{w}$ using optimization (16)
6:     Update $\mathbf{a}$ using optimization (15)
7: **end while**
8: **return**   the final architecture from learned $\mathbf{a}$.

Algorithm 2 DARTS

1: Initialize $\alpha \in \mathbb{R}^u$, $\mathbf{w} \in \mathbb{R}^n$
2: **while** not converged **do**
3:     Update $\mathbf{w}$ using optimization (4)
4:     Update α using optimization (3)
5: **end while**
6: **return**   the final architecture from learned α.

Fig. 2. An algorithm for the proposed method DARTS-CC and DARTS.

---

3   Computational Experiment

The purpose of the computational experiment is to analyze the proposed method efficiency for the neural architecture search task[1] The experiments were conducted on Fashion-MNIST [22] and CIFAR-10 [12] datasets. First, we ran an experiment to search for an architecture consisting of three cells on both datasets. We compared the proposed method with DARTS and random architecture search. After that, we ran an experiment with a large-scale architecture on CIFAR-10 and compared the results with other existing neural architecture search methods.

For the architecture search experiments, the architectures were trained for 50 epochs. The parameters of the optimization procedure were similar to those used in [6]: we used SGD for the parameter optimization with a learning rate decreasing from 0.025 to 0.001 and momentum set to 0.9. We used weight decay

---

[1] The source code for the computational experiment can be found at https://github. com/Intelligent-Systems-Phystech/DARTS-CC.

$3 \cdot 10^{-4}$ for the model parameters and 10^{-3} for the structural parameters. Similar to DARTS, after the architecture search step, we retrained the model from scratch with obtained architecture. During retraining, we also randomly dropped operations with the probability increasing from 0.0 to 0.2.

3.1 Search of the Small Architectures

During the experiments with architecture containing three cells, we did not use a *cutout* procedure: a heuristic of pruning uninformative operations from the retrained model. For all the architecture search steps, we used a batch size equal to 64. The final models were trained for 100 epochs with batch size equal to 96 and learning rate decreasing from 0.025 to 0.0.

As a set Λ we used a set of values $[10^{-10}, 10^{-6}]$, $\log_{10} \lambda \sim \mathcal{U}(-10, -6)$. for Fashion-MNIST and $[10^{-8}, 10^{-4}]$, $\log_{10} \lambda \sim \mathcal{U}(-4, -8)$. We calibrated the intervals for the λ values in order to obtain architectures with approximately similar amount of parameters to architectures obtained by DARTS. During model training we used Gumbel-Softmax distribution with temperature t decreasing from 1 to 0.2.

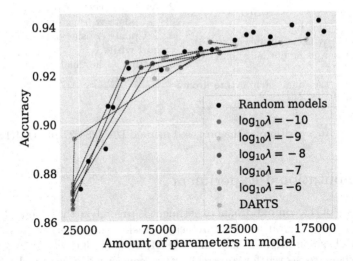

Fig. 3. The comparison of the proposed method with DARTS for the Fashion-MNIST dataset: both methods show similar performance. The proposed method allows to vary λ parameter to control the model architecture: the higher λ we use the simpler model we obtain. The lines connect models obtained from the same experiment run.

The Fig. 3 shows a dependency of the obtained models' performance on the parameter amount. Both the proposed method and DARTS gave model architectures with suboptimal performance on the Fashion-MNIST dataset. For both the proposed method and DARTS, we ran the experiment 5 times. For the proposed method we evaluated the architectures obtained from different λ values:

$\lambda \in \{10^{-10}, 10^{-9}, 10^{-8}, 10^{-7}, 10^{-6}\}$. As we can see, the proposed method gives a comparable performance to the DARTS approach. However, it allows calibrating the model complexity to find a trade-off between desired performance and model complexity.

The similar results for the CIFAR-10 dataset can be found in Fig. 4. As we can see, the proposed method and DARTS show similar performance on both datasets. The ability to control the model complexity enables us to obtain models with different structures at once and tune the final models based on the desired model complexity.

The cells obtained for different λ are shown in Fig. 5 for the Fashion-MNIST dataset. As we can see, the resulting cells tend to have a less complex structure with increasing λ parameter.

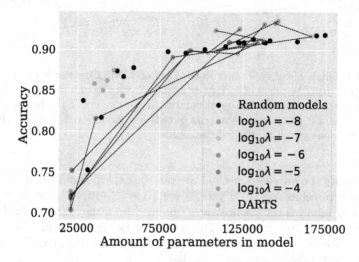

Fig. 4. The comparison of the proposed method with DARTS for the CIFAR-10 dataset.

3.2   Large-Scale Architecture Search

In order to compare the proposed method with different neural architecture search methods, we conducted an additional experiment on the CIFAR-10 dataset. For this experiment, we used settings similar to those described in [6]. We used a model with eight cells during architecture search and with 20 cells during the retraining model from scratch. The final models were trained for 600 epochs with batch size equal to 72 and learning rate decreasing from 0.025 to 0.0.

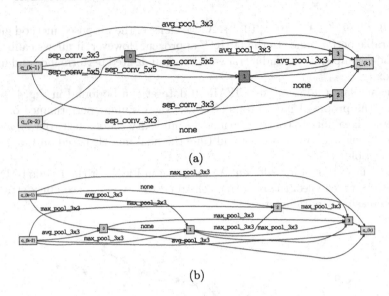

(a)

(b)

Fig. 5. An example of cells obtained on the Fashion-MNIST dataset for $\lambda = 10^{-10}$ (a) and $\lambda = 10^{-6}$ (b). The colors of each node in the cell vary from blue to red. The closer to the red color, the greater the total number of parameters of the incoming edges. Here c_{k-1} and c_{k-2} are the outputs of cells $k - 1$ and $k - 2$, respectively, c_{k} is the output node of cell k. (Color figure online)

To evaluate the proposed method, we ran the experiment 3 times. As a set Λ we used a set of values $\lambda \in [10^{-10}, 10^{-6}]$, $\log_{10} \lambda \sim \mathcal{U}(-10, -6)$. For the proposed method, we evaluated the architectures obtained from the highest and the lower values of $\lambda : \lambda \in \{10^{-10}, 10^{-6}\}$. The results of the experiments are shown in Table 1. As we can see, the resulting models achieve a performance comparable to other neural architecture search methods.

Based on the obtained results, we can conclude that the proposed method allows us to obtain models with different model complexity per one architecture search optimization procedure. The resulting architectures give us a rather good performance comparable to architectures of other neural architecture search methods. At the same time, our model enables us to control the model complexity and obtain different models based on the desired computational budget.

Table 1. Top-1 Accuracy on CIFAR-10 dataset. The baseline results are taken from [20].

Architecture	Accuracy (%)	Params(M)
DenseNet-BC [9]	96.54	25.6
NASNetA + cutout [25]	97.35	3.3
AmoebaNet-B + cutout [19]	97.87	34.9
PNAS [15]	96.59 ± 0.09	3.2
NAONet [17]	96.82	10.6
SMASHv2 [2]	95.97	16
SETN + cutout [5]	97.31	4.6
GDAS + cutout [6]	96.25	2.5
DARTS(2nd order) + cutout [16]	97.24 ± 0.09	3.3
SNAS (mild) + cutout [23]	97.02	2.9
PR-DARTS DL1 + cutout [13]	97.26 ± 0.12	3.2
PC-DARTS + cutout [24]	97.43	3.6
PDARTS + cutout [4]	97.50	3.4
Amended-DARTS S1 + cutout [1]	97.19 ± 0.21	3.5
DARTS+ with cutout [14]	97.68	3.7
DARTS-CC + cutout, $\lambda = 10^{-10}$	97.13 ± 0.08	4.4 ± 0.5
DARTS-CC + cutout, $\lambda = 10^{-6}$	96.58 ± 0.38	2.2 ± 0.2

4 Conclusion

In this paper, we considered the problem of deep learning model selection with model complexity control. To control the model complexity, we proposed a modification of the differential architecture search algorithm called DARTS-CC. We considered the structural parameters of the architecture as a function of the model complexity parameter. To evaluate the performance of the proposed method, we conducted experiments on the Fashion-MNIST and CIFAR-10 datasets and compared the obtained architecture with architectures gathered from the DARTS method. The results showed the comparable performance of the method. The proposed method allows us to control the trade-off between model characteristics: the model performance and the model complexity. Future research plans include researching the robustness of the obtained architecture and properties of the proposed optimization problem.

References

1. Bi, K., Hu, C., Xie, L., Chen, X., Wei, L., Tian, Q.: Stabilizing darts with amended gradient estimation on architectural parameters. arXiv preprint arXiv:1910.11831 (2019)

2. Brock, A., Lim, T., Ritchie, J.M., Weston, N.: Smash: one-shot model architecture search through hypernetworks. arXiv preprint arXiv:1708.05344 (2017)
3. Chang, J., Zhang, X., Guo, Y., Meng, G., Xiang, S., Pan, C.: Differentiable architecture search with ensemble Gumbel-Softmax. arXiv preprint arXiv:1905.01786 (2019)
4. Chen, X., Xie, L., Wu, J., Tian, Q.: Progressive differentiable architecture search: bridging the depth gap between search and evaluation. In: Proceedings of the IEEE/CVF International Conference on Computer Vision, pp. 1294–1303 (2019)
5. Dong, X., Yang, Y.: One-shot neural architecture search via self-evaluated template network. In: Proceedings of the IEEE/CVF International Conference on Computer Vision, pp. 3681–3690 (2019)
6. Dong, X., Yang, Y.: Searching for a robust neural architecture in four GPU hours. In: Proceedings of the IEEE/CVF Conference on Computer Vision and Pattern Recognition, pp. 1761–1770 (2019)
7. Grebenkova, O., Bakhteev, O.Y., Strijov, V.: Variational deep learning model optimization with complexity control. Informatika i Ee Primeneniya (Inform. Appl.) **15**(1), 42–49 (2021)
8. Ha, D., Dai, A.M., Le, Q.V.: Hypernetworks. CoRR abs/1609.09106 (2016). http://arxiv.org/abs/1609.09106
9. Huang, G., Liu, Z., Van Der Maaten, L., Weinberger, K.Q.: Densely connected convolutional networks. In: Proceedings of the IEEE Conference on Computer Vision and Pattern Recognition, pp. 4700–4708 (2017)
10. Jang, E., Gu, S., Poole, B.: Categorical reparameterization with Gumbel-Softmax. arXiv preprint arXiv:1611.01144 (2016)
11. Jin, X., et al.: RC-Darts: resource constrained differentiable architecture search. CoRR abs/1912.12814 (2019). http://arxiv.org/abs/1912.12814
12. Krizhevsky, A., Hinton, G., et al.: Learning multiple layers of features from tiny images (2009)
13. Laube, K.A., Zell, A.: Prune and replace NAS. In: 2019 18th IEEE International Conference On Machine Learning and Applications (ICMLA), pp. 915–921. IEEE (2019)
14. Liang, H., et al.: DARTS+: improved differentiable architecture search with early stopping. arXiv preprint arXiv:1909.06035 (2019)
15. Liu, C., et al.: Progressive neural architecture search. In: Ferrari, V., Hebert, M., Sminchisescu, C., Weiss, Y. (eds.) ECCV 2018. LNCS, vol. 11205, pp. 19–35. Springer, Cham (2018). https://doi.org/10.1007/978-3-030-01246-5_2
16. Liu, H., Simonyan, K., Yang, Y.: DARTS: differentiable architecture search. CoRR abs/1806.09055 (2018). http://arxiv.org/abs/1806.09055
17. Luo, R., Tian, F., Qin, T., Chen, E., Liu, T.Y.: Neural architecture optimization. arXiv preprint arXiv:1808.07233 (2018)
18. Maddison, C.J., Mnih, A., Teh, Y.W.: The concrete distribution: a continuous relaxation of discrete random variables. arXiv preprint arXiv:1611.00712 (2016)
19. Real, E., Aggarwal, A., Huang, Y., Le, Q.V.: Regularized evolution for image classifier architecture search. In: Proceedings of the AAAI Conference on Artificial Intelligence, vol. 33, pp. 4780–4789 (2019)
20. Tanveer, M.S., Khan, M.U.K., Kyung, C.M.: Fine-tuning DARTS for image classification. In: 2020 25th International Conference on Pattern Recognition (ICPR), pp. 4789–4796. IEEE (2021)
21. Wu, B., et al.: FBNet: hardware-aware efficient convnet design via differentiable neural architecture search. In: Proceedings of the IEEE/CVF Conference on Computer Vision and Pattern Recognition, pp. 10734–10742 (2019)

22. Xiao, H., Rasul, K., Vollgraf, R.: Fashion-MNIST: a novel image dataset for benchmarking machine learning algorithms. arXiv preprint arXiv:1708.07747 (2017)
23. Xie, S., Zheng, H., Liu, C., Lin, L.: SNAS: stochastic neural architecture search. arXiv preprint arXiv:1812.09926 (2018)
24. Xu, Y., et al.: PC-DARTS: partial channel connections for memory-efficient architecture search. arXiv preprint arXiv:1907.05737 (2019)
25. Zoph, B., Vasudevan, V., Shlens, J., Le, Q.V.: Learning transferable architectures for scalable image recognition. In: Proceedings of the IEEE Conference on Computer Vision and Pattern Recognition, pp. 8697–8710 (2018)

Author Index

Printed in the United States
by Baker & Taylor Publisher Services